D0856238

# Entomology
## High-school science fair projects

H. Steven Dashefsky
Illustrations by Janice Lebeyka

TAB Books
Division of McGraw-Hill, Inc.
Blue Ridge Summit, PA 17294-0850

FIRST EDITION
FIRST PRINTING



**Library of Congress Cataloging-in-Publication Data**

Dashefsky, H. Steve.
Entomology : high-school science fair experiments / by H. Steven Dashefsky.
p. cm.
Includes bibliographical references and index.
ISBN 0-07-015661-1 (h) ISBN 0-07-015662-X (p)
1. Entomology projects. I. Title. II. Title: High-school science fair experiments.
QL468.5.D36 1993
595.7'0078--dc20 93-39693
CIP

Acquistions editor: Kim Tabor
Editorial team: Joanne M. Slike, Executive Editor
Vince Peciulis, Editor
Joann Woy, Indexer
Production team: Katherine G. Brown, Director
Rose McFarland, Layout
Tina M. Sourbier, Typesetting
Susan E. Hansford
Tara Ernst, Proofreading
Design team: Jaclyn J. Boone, Designer
Brian Allison, Associate Designer
Cover design: Sandra Blair Design, Harrisburg, Pa.
Cover photograph: Bender and Bender Photography, Waldo, Ohio
Cover photograph is a closeup shot of a ladybug's shell.
TAB2
4468

# Contents

PART II
## INSECT LIVES



PART III
## ENVIRONMENTAL ISSUES



PART IV
## INSECT ECOLOGY



PART V
## INSECT BEHAVIOR

*For Lindsay, my budding*
*naturalist and favorite hiking partner*

# Disclaimer

Adult supervision is required when working on these projects. No responsibility is implied or taken for anyone who sustains injuries as a result of using the materials or ideas, or performing the procedures put forth in this book. Use proper equipment (gloves, forceps, safety glasses, and so on) and take other safety precautions such as tying up loose hair and clothing and washing your hands when the work is done. Use chemicals, dry ice, boiling water, flames, or any heating elements with extra care. Hazardous chemicals and live cultures (organisms) must be handled and disposed of according to appropriate directions set forth by your sponsor. Follow your science fair's rules and regulations and any standard scientific practices and procedures set forth by your school or other governing body.

Additional safety precautions and warnings are mentioned throughout the text. If you use common sense and make safety a first consideration, you will create a safe, fun, educational, and rewarding project.

# Acknowledgments

My thanks to Dr. Paula J.S. Martin, Carolyn Bardwell, and Vincent D'Amico for their help devising many of these projects and for their technical assistance performing them. I also want to thank Allissa Brotman for her award-winning science fair project about the effect of electromagnetic radiation on fruit flies (Project 8).

Finally, my thanks to Dr. John Edman, Dr. John Clark, and Dr. John Stoffolano at the University of Massachusetts, Department of Entomology, for their continuing support.

# A word about safety and supervision

All the projects in this book require an adult sponsor to ensure the student's safety and the safety of others. Science Service, Inc. is an organization that sets science fair rules, regulations, and safety guidelines and holds the International Science and Engineering Fairs (ISEF). This book recommends that the student who performs the projects follow the ISEF guidelines pertaining to adult supervision—and it is assumed that the student will do so. The ISEF guidelines state that any student undertaking a science fair project should have an adult sponsor assigned to him or her.

The adult sponsor is described as a teacher, parent, professor, or scientist in whose lab the student is working (for the purpose of this book, the sponsor will usually be the student's teacher). The adult sponsor must have a solid background in science and must be in close contact with the student throughout the project. The sponsor is responsible for the safety of the student who is conducting the research, which includes the handling of all equipment, chemicals, and organisms. The sponsor must also be familiar with the regulations and commonly approved practices that govern chemical and equipment usage, experimental techniques, the use of laboratory animals, cultures and microorganisms, and proper disposal techniques.

If the adult sponsor is not qualified to handle all of these responsibilities, the sponsor must assign the responsibilities to someone who is capable of handling them. Before proceeding with a project, most science fairs require that the adult sponsor fill out appropriate forms identifying the sponsor's qualifications.

The sponsor is responsible for reviewing the student's Research Plan, described later in this book, and for making sure that the experiments are done according to local, federal, and ISEF (or other appropriate governing body's) guidelines.

Before beginning, the student and the adult sponsor should read and review the entire project. The adult should determine which portions of the experiment can be performed without supervision and which portions will require supervision. In addition, ⚠ symbols thoughout the text indicate where extra caution, such as wearing safety goggles and gloves must be taken.

For a copy of the ISEF's rules and regulations, contact Science Service, Inc., at 1719 N Street, N.W., Washington, DC 20036, (202) 785-2255. This book includes a checklist for the adult sponsor, approval forms, and valuable information on all aspects of participating in a science fair.

# How to use this book

There are two ways to use this book. If you are new to science fair projects and feel that you need a great deal of technical guidance, you can use the projects as explained with few if any adjustments. These are good, solid science fair projects. However, if you want to be a contender for an award-winning project, you must use the experiments in the book as the foundation—and then read and incorporate into the project additional ideas. Many suggestions are given in the "Going Further" or "Suggested Research" sections of each project. This book is designed to provide a core experiment with many suggestions about how to expand the scope or adjust the focus of the experiment.

As explained in the section on "Scientific Research," every experiment builds on what was learned from a previous experiment, thus advancing science each step of the way. Many projects in this book consist of more than one experiment. One experiment advances the other or in some cases, confirms the other.

Each project in this book has the following sections:

- Background
- Project Overview
- Materials List
- Procedures

- Analysis
- Going Further
- Suggested Research.

## Background

This section provides background information about the topic to be investigated. It offers you a "frame of reference" so that you can see the importance of the topic and why research is necessary to advance our understanding. This section could be considered the initial step in your literature search. (See the next section for more about your literature search.) Although a small step, it is enough to see if the subject piques your interest.

## Project Overview

If the Background section interests you, continue reading because the Project Overview section describes the purpose of the project. It explains the problems that exist and poses questions that the experiment is intended to resolve. These questions can be used to formulate your hypothesis. An illustration often shows the apparatus setup, if one exists, so that you can see what to expect. Be sure to discuss this section, as well as the next with your sponsor to see if the requirements can realistically be met.

## Materials List

The Materials List section gives everything needed to perform the experiment. Be sure you have access to or can get everything before beginning. Some equipment or apparatus are expensive. Check with your teacher to see if all the equipment is available in your school or can be borrowed from elsewhere. Be sure your budget can handle anything that must be purchased. A list of many scientific supply houses is provided at the back of the book. (Part numbers are supplied for Ward's Natural Sciences and are listed by chapter number in the back of this book).

Although most people don't think of research scientists as a group of hammer-and-nails carpenters, frequently they must be. Building a device or experimental workstation often involves many trips to the hardware store for supplies, a little sweat on the brow, and a lot of ingenuity.

Living organisms, such as bacterial cultures, probably must be ordered from the supply houses. Others, such as insects, can be ordered, purchased locally, or caught—depending on the project, your location, and the time of year. If you are using live organisms, work with your sponsor to be sure that you adhere to all science fair regulations and standard biological research practices. Before beginning, discuss with your sponsor the proper way to dispose of any hazardous materials, chemicals, or cultures. (See Fig. I-12.)

## Procedures

The Procedure section gives step-by-step instructions on how to perform the experiment and suggestions on how to collect data. Be sure to read through this section with your sponsor before undertaking the project. Illustrations are often used to clarify procedures. Although each step is given, some projects require standard procedures such as inoculating a petri dish. These steps are often stated but not explained. Your sponsor can help you with these standard procedures.

## Analysis

The Analysis section doesn't draw conclusions for you. Instead, it asks questions to help you analyze and interpret the data so that you can come to your own conclusions. In many cases, empty tables and charts are provided for you to begin your data collection. You should convert as much of your raw data as possible into line and bar graphs or pie charts.

Some experiments might require statistical analysis to determine if there are significant differences between the experimental groups and the control group. Check with your sponsor to see if you should perform statistical analysis for your project and, if so, what kind. See the "Suggested Reading" section for books that will help you analyze your data.

## Going Further

This section is a vital part of every project. It lists many ways for you to continue researching the topic beyond the original experiment. These suggestions can be followed as is or, even more importantly, they might spark your imagination to think of some new twist or angle to take while performing the project. These suggestions may show ways to more thoroughly cover the subject matter or may show you how to broaden the scope of the project. The best way to assure an interesting and fully developed project is to include one or more of the suggestions from the "Going Further" section or include an idea of your own that was inspired from this section.

## Suggested Research

This section suggests new directions to follow while researching the project. It often suggests reading material and lists sources of additional information. In addition to written materials, it often includes organizations, companies, or other sources of information. Using these additional resources might turn your project into a winner.

# PART I

# Before You Begin

Before delving into any scientific experiment, there are three things to understand: the terminology used, the methodology required, and the suitability of the project to your own situation and preferences. The following three chapters examine these elements.

# 1

# An introduction to entomology

There are more species of insects on our planet than all other organisms combined. In fact, roughly three out of every four identified species are insects, and most scientists believe that we have only identified a fraction of the total. Today, roughly 800,000 types of insects have been identified, but estimates of how many actually exist range from 5 million to 30 million. Even more amazing than the number of different types of insects is the number of any one kind of insect. One female fruit fly in May can produce millions of fruit flies by September.

The importance of insects on our planet is hard to comprehend. Most food webs would collapse if insects were eliminated from their intricate and complex set of relationships. They not only are important consumers in most grazing food chains but also are vital to most detritus (decomposing) food chains that return nutrients back to the earth.

## For Better and for Worse

Insects provide us either directly or indirectly with food and other necessities of life. Many food and feed crops would cease to exist without insect pollination. Many vegetables and most fruits that we eat exist because of insect pollination. (See FIG. 1-1.) The fact is that life, as we know it, could not exist on this planet without these legions of insects.

Throughout history, however the human race, has spent a great deal of its time and energy fending off insects. Even though only a small percentage of insects are considered pests, they continually threaten the world's food supply. In many parts of the world, far more of our crops end up in the mouths of insects than in our own. Our modern, high-tech

**Fig. 1-1** *Pollinating plants is one of the many important roles insects play in an ecosystem.*

farms—on which a single crop is grown over vast expanses of land (called *monocultures*)—make the situation worse by providing virtually unlimited resources to expanding insect populations.

Other insects are involved in the transmission of diseases, which through the ages have killed large portions of the human race. During the fourteenth century, fleas helped spread a plague that killed 25 million people thoughout Europe—25% of-Europe's population. (See FIG. 1-2.) Malaria, yellow fever, and dengue fever are spread by mosquitoes, and sleeping fever is spread by the tsetse fly. These and many other scourges are often ignored in more developed nations but still devastate less developed regions of the world.

In our battle to conquer these insect enemies, synthetic pesticides have become our primary weapon on the field of combat. Our entire planet is becoming contaminated because of the excessive application of pesticides

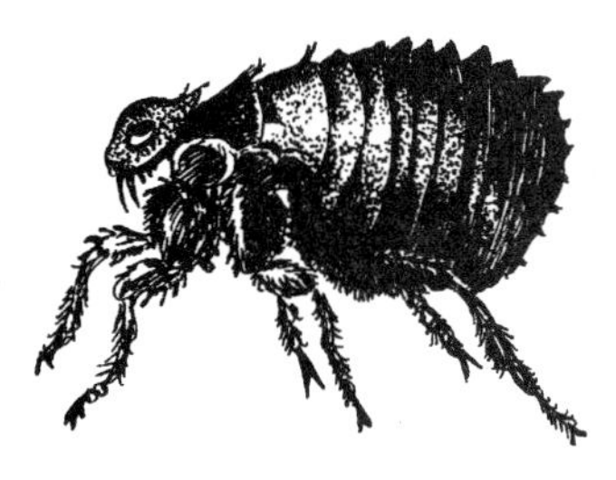

**Fig. 1-2** *Some insects, such as the flea, can spread diseases.*

used to protect our crops from these insect pests. Pesticide residues are found in our drinking water, food supply, and even in some mothers' breast milk. The amazing diversity of insects is demonstrated by the fact that the best alternative to using pesticides is the use of beneficial insects.

## Defining the Study of Insects

*Entomology* is a term that refers to the study of insects—not only pure scientific research, such as studying dragonfly flight dynamics or migration patterns of Monarch butterflies, but also applied research, such as identifying predator-prey relationships, or hormones and pheromones for pest-control applications.

Scientific research often uses insects as research subjects. For example, genetic studies historically have used insects because of insects' rapid life cycles and reproductive capabilities, while ecological studies have often used insects because they are good indicators of an ecosystem's health. Projects in this book include research that specifically studies insects as well as research that uses insects as part of a study.

Most science fairs require that projects be placed into specific categories for purposes of judging. The ISEF uses 14 broad categories. Almost all of the projects in this book fall into the "Zoology" category. A few could be placed, however, into the "Environmental Science" category.

## Things to Know

For anyone interested in the study of insects, there are some things that you should know. All of the items mentioned below are not needed for all the projects in this book. Many items, however, such as using a field guide, will be useful in your research or will help to enhance your project—for example, by adding a specialized insect collection to your project.

### An insect field guide

Before pursuing any study of insects, it's an excellent idea to get an insect field guide. There is a list of these guides at the back of this book. Many of these guides probably can be found in your local or school library or can be purchased in a book, science, or nature store. Field guides do more than just match pictures and names of insects. The guides explain the traits that distinguish insects from each other, explain insects' life cycles and behaviors and usually describe how to collect, preserve, and rear insects.

### Identifying insects

Once you have an insect field guide, try to become familiar with the proper names of insects. Every experiment involving live animals should include not only the correct scientific name of each organism involved but also an explanation of how that organism fits in with similar organisms. With insects, using proper names is especially important because there are so many kinds of insects and so much confusion about their names.

Even though scientific names sound difficult, they never change, and anyone in any country can tell exactly what insect you are referring to when the scientific name is used. Common or everyday insect names, on the other hand, are often confusing. For example, dragonflies and damselflies are not really "flies," since they belong to the order *Odonata*, not *Diptera* (which are the true flies). Furthermore, all insects are not "bugs," since only true bugs belong to the order *Hemiptera*. And all insects are not "beetles" because real beetles belong to the order *Coleoptera*. (See FIG. 1-3.)

**Fig. 1-3** *Commonly called a ladybug, this insect is actually a beetle.*

When you research an insect, the best way to be sure that you are gathering information about the correct insect is to use its scientific name. Consulting your field guide is the best way to become familiar with proper insect identification and scientific names. Using these names should not pose a problem since most projects in this book only deal with one or two insects.

## Making a collection

An insect collection is a great part of any science fair project that involves insects. All of the experiments in this book would benefit from a small collection of the insects involved. The collection could simply consist of samples of the insects used in the project, a survey of the insects found in a particular ecosystem, or possibly the life cycle of the insect used as a research specimen.

*Collecting gear* If you decide to include a collection in your project, you'll need some collecting gear. First, be sure the field guide that you are using includes information on making a collection.

You'll need an insect-collecting net. Depending on the project you select, you'll need a butterfly net for capturing insects in air, a sweep net for gathering insects in vegetation, or an aquatic net for catching insects in or on water. These nets can be purchased in science or nature stores (for example, such stores as The World of Science and The Nature Company, which can be found in shopping malls) or from a scientific supply house, but you might be able to borrow one from your school. There are dual-purpose nets that do well for both aerial and sweep situations. In most cases large, fine sieves or strainers can be used instead of aquatic nets. (Scientific supply houses are listed at the back of this book.)

After collecting insects, you'll need an insect "killing jar," which can be made or purchased from a supply house. A *killing jar* is a container, such as a mayonnaise jar, with a lid that can close tightly and a substance at the bottom, such as plaster of paris, that can absorb fluid. An *activating fluid* is soaked into the plaster of paris. This activating fluid produces fumes that kill the insects. Nail polish remover can be used as an activating fluid, but ethyl acetate is preferred. Ethyl acetate is available from scientific supply houses but may be available at your school.

CAUTION: Follow the handling instructions that accompany the activating fluid, or ask your sponsor for directions. It is dangerous (very flammable) and must be handled with caution. Never inhale the fumes. (See FIG. 1-4.)

**Fig. 1-4** *A "killing jar" usually contains ethyl acetate, which kills insects to be studied or preserved.*

If you want to build your own jar, follow these instructions. Fill the bottom of a mayonnaise jar with about 1 inch of plaster of paris and add water according to the instructions for plaster. When the plaster is completely dry, pour into the jar a small amount of ethyl acetate, which will act as the activating fluid. This fluid should be completely absorbed by the plaster of paris. The fluid must be replenished occasionally, since it evaporates each time that the jar is opened. The more the jar is opened, the sooner it must be replenished with fluid.

*Collecting the insects* Large numbers of small insects can be collected rapidly with a sweep net in tall vegetation, such as an overgrown field. To "sweep net," move the net rapidly back and forth through the grass in front of you while walking slowly through the field. (See FIG. 1-5.) The net should be hitting the top few inches of the vegetation. Walk and sweep in this manner for about 5 minutes. To stop, flip the net closed by twisting the net opening downwards. (See FIG. 1-6.) This procedure folds the net over

**Fig. 1-5** *Sweep the open field that is adjacent to a wooded area to see the difference in biodiversity.*

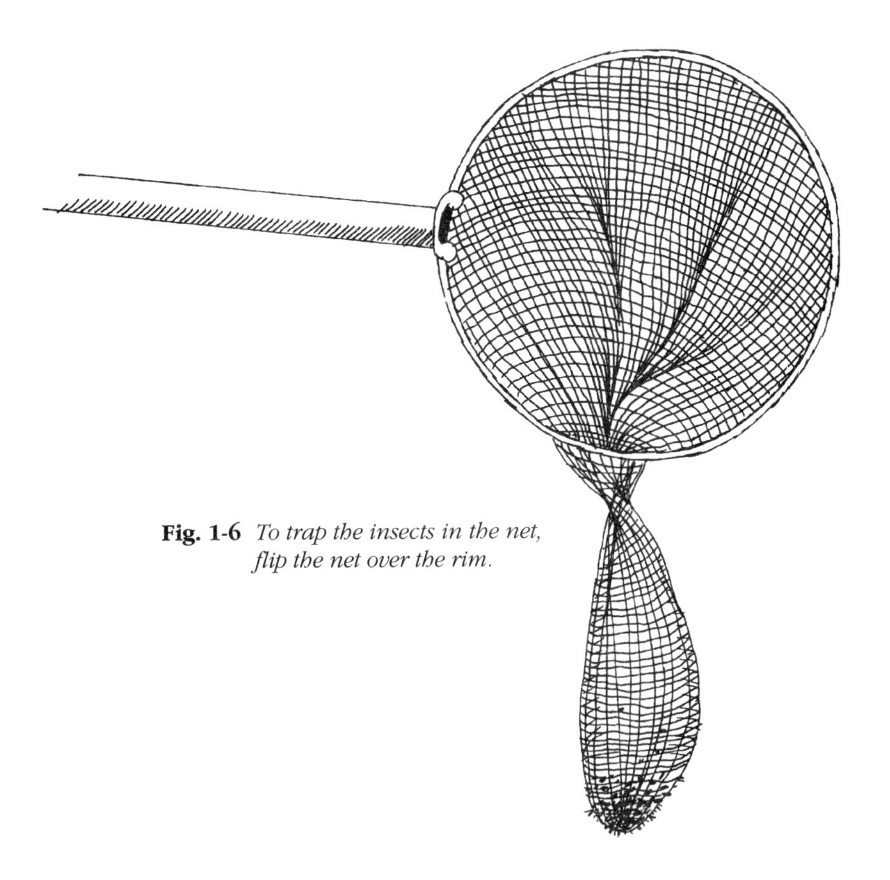

**Fig. 1-6** *To trap the insects in the net, flip the net over the rim.*

the net rim and traps the insects inside. Most of the insects will be in the bottom of the net. Stick the bottom portion of the net (containing the insects) into the killing jar and hold the jar cap over the top. (You won't be able to close the cap since the remainder of the net will be out of the jar.) After a few minutes the insects will be incapacited. (See FIG. 1-7.) Remove the net from the jar and dump the insects out of the net into the killing jar. Seal the jar tightly and leave the insects in the jar at least one hour.

Very small flying insects can be difficult to catch, so consider making (or purchasing) a *collecting aspirator*, which is simply a glass vial with a two-hole rubber stopper at the end. An angled glass tube sticks out of one hole and a plastic hose comes out of the other hole. You place the glass tube near the insect to be collected and suck on the hose at the other end. This sucks the insect into the cylinder. A fine net at the end of the tube assures that you don't accidentally suck the insect through the tube and into your mouth. (See FIG. 1-8.)

With a net, large flying insects can be collected individually. You can also use tweezers or forceps to pick up small insects that you find in wood, leaf litter, or a stream or pond. Insects can be found almost anywhere. Look in your field guide for more suggestions about how to collect insects.

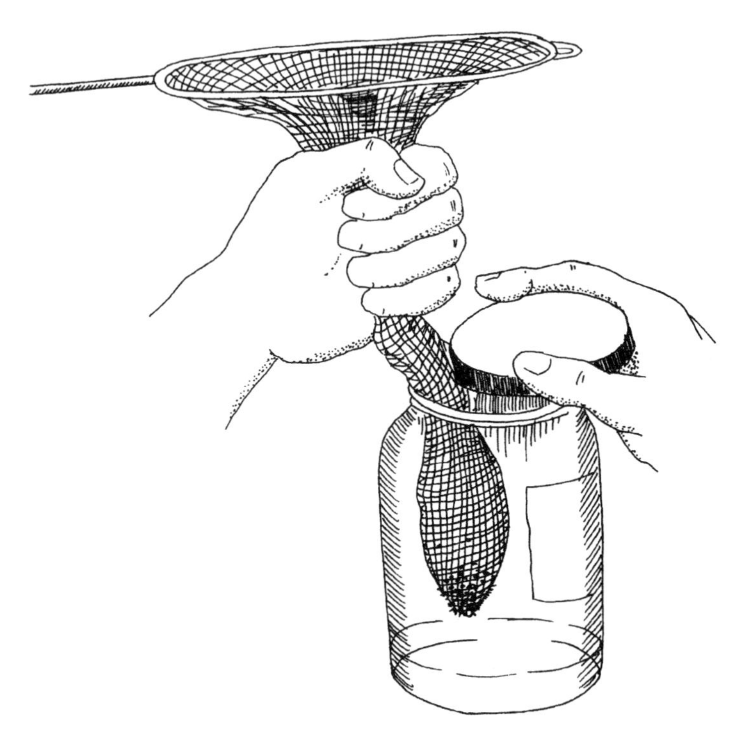

**Fig. 1-7** *To temporarily incapacitate the insects, hold the bottom portion of the net in the jar and hold the cap in place.*

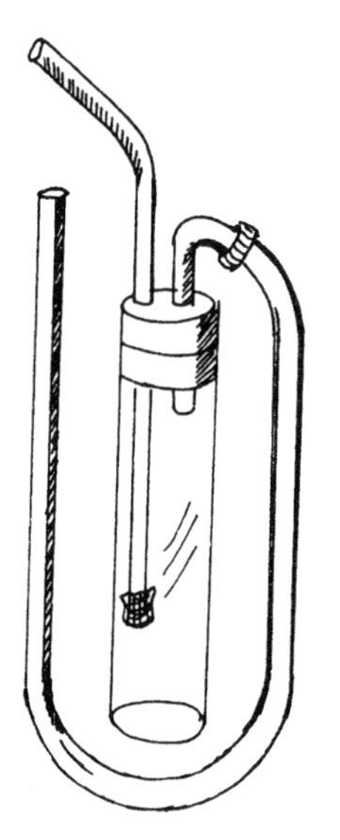

**Fig. 1-8** *Using a collecting aspirator is an easy way to catch small insects.*

*Preserving and displaying your collection* How do you display the insects you've caught? Insects can be pinned and stored in a box (similar to a cigar box), or they can be placed in Riker Mounts, which are glass covered boxes filled with cotton. The cotton protects the insects, and the glass cover lets you view the insects within. Soft-bodied insects can be placed in vials filled with 70% rubbing alcohol so that they won't deteriorate. See your insect field guide for details on collecting, pinning, and mounting insects.

## Handling & Rearing Live Insects

Raising insects and watching their development is a great way to study not only insect biology but also insect behavior. The first requirement for such study is a container to hold the insects. You can use any tightly sealed container with an opening for air. Glass jars work well because you can watch the insects through the glass. The air opening must be screened so that the insects can't escape. Nylon material (such as pantyhose) works well as a screen and should be secured over the jar opening with a rubberband. Window screening is sturdier, but very small insects (such as fruit flies) could escape.

Like all animals, insects need food and water. A good way to supply water is to soak a cotton ball in water and place it on a surface that does not absorb water (like a plastic bag). The insect can drink the water from the cotton. A bowl of water would probably drown the insects.

Different types of insects eat different types of foods. Plant-eating insects should have a supply of their host plant (the plant that they regularly feed on). If you plan to raise plant-eating insects, gather some leaves and twigs from the host plant when you collect those insects. Be sure to remember the plant's location since you might need to gather more of that plant. Predatory insects (insects that eat other animals, usually other insects) must have a supply of their live prey (the insects that they feed on).

In transferring small, live insects into or out of a cage, it is hard to hold them with your hands, or even with a pair of forceps, without damaging them. The solution is a "transfer aspirator," which is different from a "collecting aspirator" (mentioned earlier). (See FIG. 1-9.) A "transfer aspirator" is simply two cylinders that you suck the insects into and blow them out of. The two cylinders are held together with plastic tubing. A mesh net at the end of one tube prevents the insects from being sucked into your mouth. Transfer aspirators can be made or purchased.

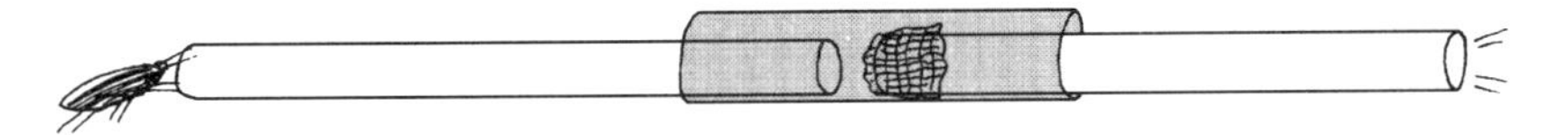

**Fig. 1-9** *The transfer aspirator allows you to collect and relocate very small insects.*

Even though insects are all around us and are a part of our daily lives, care must be taken when observing—and, especially, when handling—insects. Don't assume that an insect is harmless. Follow all instructions and heed all warnings in this book, especially those that pertain to the handling of insects.

## Special considerations

Some projects must be done at a certain time of year. For example, aphids are not available during the winter and cannot be ordered from supply houses. You can use your imagination and check greenhouses or florists, but don't proceed with an experiment until you have considered the time of year and availability of insects. The "Materials List" section in each project describes the live insects that are needed and gives suggestions about where you can get them. The insects for many of the experiments can be caught, but some can be purchased in a pet or bait store or from a supply house. In a few experiments, the live insects must be purchased from a supply house, since they won't be available any other way. (A list of supply houses with their telephone numbers is given at the end of this book.)

Some projects require that you raise insects demanding considerable attention. Check with your sponsor to be sure that you have the time and the resources to complete any project that you begin.

# 2

# An introduction to scientific research

Science fairs give you the opportunity not only to learn about a topic but also to participate in the discovery process. Although you probably won't discover something previously unknown to mankind (although you never can tell), you will perform the same procedure—and experience the same process—by which scientific discoveries are made. Advances in science move forward slowly, with each experiment building upon a previous one and preparing researchers for the next. Advances in medicine, biotechnology, agriculture—and virtually all scientific disciplines—proceed one step at a time. A typical science-fair project should allow you to see what it is like to take one or two of these steps for yourself. The following example typifies how science moves forward.

A problem such as controlling a particular type of insect pest without harmful pesticides might be solved by a series of scientific experiments. First, field studies could search for natural predators and parasites of the insect. Lab and field studies could then be performed to determine how effective each natural enemy could be at controlling the insect. (See FIG. 2-1.) The life cycles of these insects would have to be studied as well as the way each type of insect fits into the local ecosystem.

Further studies might be performed to determine the population dynamics of these natural enemies of the pest. Are they capable of controlling the pest? What would happen to the entire ecosystem if the number of these predators dramatically increased by artificial means? Experiments might find that some of the pest's enemies are incapable of controlling the insect. Those experiments would still be valuable, however, since they would provide information that would keep scientists on the correct track.

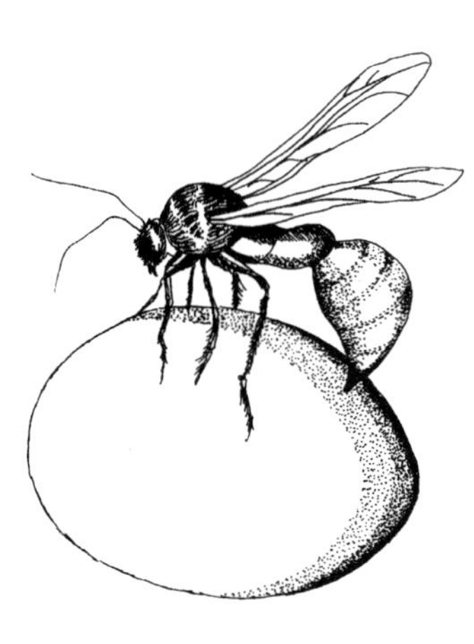

**Fig. 2-1** *Biological control methods use beneficial insects such as this tiny wasp to control insect pests.*

Once a natural enemy is discovered to be a likely candidate, small-scale testing can begin until a solution to the problem is found.

As you can see in the example above, each experiment was necessary before the next could be performed, and the entire progression was necessary before a successful conclusion could be reached. Most scientific research, no matter how simple or how sophisticated, follows a protocol that demands consistency and, most importantly, the ability to repeat an experiment to see if the results would be the same. When one scientist or research team finds some new revelation, others must verify it. The scientific method provides a framework for researchers to follow. It assures a highly focused, reproducible sequence of events. The basics of the scientific method are discussed below.

## The Scientific Method

The scientific method can be divided into five steps. The paragraphs below describe each step and relate them to sections in this book.

### Problem

What question do you want to answer, or what problem would you like to solve? For example, how do ants warn their neighbors of danger? Can a certain type of insect be used to control a particular insect pest? The "Project Overview" section of each project in this book, gives a number of questions and problems to think about. The "Going Further" and "Suggested Research" sections in this book can also give you ideas for sparking your imagination about other problems.

### Hypothesis

The *hypothesis* is an educated guess, based on your literature search, that offers a possible answer to the questions posed. For example, you might hypothesize that sound, sight, or chemicals are used by ants to warn their neighbors of danger or that a certain type of parasite can control a particular insect pest. You can form a hypothesis about any or all of the questions given in the "Project Overview" section.

## Experimentation

The experiment is designed to determine whether the hypothesis is correct or not. Even if the hypothesis isn't correct, a well designed experiment helps determine why it isn't.

There are two major parts to an experiment. The first part involves designing and setting up the experiment. You should ask such questions as: What preparations and procedures must be followed to test the hypothesis? What materials will be needed? What live organisms, if any, are needed? What step-by-step procedures must be followed during the experiment? What observations and data must be made and collected while the experiment is running? When these questions have been answered, the actual experiment can be performed.

The second part involves performing the experiment, making observations, and collecting data. The results must be documented for study and analysis. The more details, the better. There are three important things to remember when performing research: Take notes, take notes and take notes. (See FIG. 2-2.) The most common mistake that new scientists make is thinking that they will remember some minute obser-

**Fig. 2-2** *Nothing is more important as you prepare your science project than taking detailed notes.*

vation. If you always carry a notebook and pencil when working on your project, remembering won't be a problem. Some science fairs require that the project notebook be submitted along with a brief abstract of the project. Some fairs require or encourage a full-length report of the project as well.

The "Materials" section of each project lists all the materials needed for each experiment, and the "Procedures" section gives step-by-step instructions. Suggestions are given about what observations should be made and what data should be collected.

Replication is another important aspect of experimentation. For any project to be considered valid scientific work, the experiments in each group should be replicated or repeated as many times as possible. The groups of replicated experiments can then be averaged together—or, better yet, statistically analyzed. For the projects in this book, try to perform each of the experiment groups in triplicate. For example, if you are collecting samples from a site, collect three times in the general area, or if you are culturing organisms, establish three such cultures. Replication reduces the chances of collecting spurious data, which will result in erroneous analysis and conclusions.

### Analysis

Once you have completed the experiment and have collected the data, you must analyze it and draw conclusions in order to determine if your hypothesis is correct. You may create tables, charts, or graphs to help analyze the data. The "Procedures" section of each project suggests what observations to make and what data to collect while running the experiment. The "Analysis" section asks important questions to help you analyze the data and often contains empty tables or charts to fill in with your data. This book provides guidance, but you must draw your own conclusions. The "Suggested Reading" section of this book lists other books that can help you analyze data.

The conclusions should be based upon your original hypothesis. Was it correct? Even if it was incorrect, what did you learn from the experiment? What new hypothesis can you create and test? Something is always learned while performing an experiment, even if it's how *not* to perform the next experiment.

## Building on Past Science Fair Projects

Just as scientists advance the work of other scientists, so too can you advance the work of those that have performed other science fair projects before you. I don't mean copying their work, but thinking of the next logical step in that line of research. Possibly you can put a new twist on a previous experiment by testing other hypotheses. For example, if electromagnetic radiation can damage aquatic plant life, can it harm the microbes that live among the plants? Or, if an original experiment was performed in vitro (in a test tube), can a similar experiment be performed in vivo (in nature)?

Abstracts of previous science fair projects are available from the Science Service in Washington, DC. See the "Reference" section in this book for sources of successful science fair projects. (See also the "Selecting a Project" section.)

# 3

# Selecting a project & getting started

You are probably looking through this book because you're interested in insects. Therefore, the first thing to do is to find out what specifically piques your interest, if you don't already know. There are a few ways to find out.

## Exploring

Begin exploring the world of insects by simply watching them. Start by looking for them in the world around you: in your home, your backyard, the overgrown field down the street, or the cracks in the pavement. (See FIG. 3-1.)

Watch the insect's behavior. See how it finds food, acts with other insects of the same kind or when other types of insects are around. Look for ways in which the insect affects its environment and the environment affects the insect. Look at the insect's body structure or investigate its habitat. If you find yourself saying that you'd "like to know more about" something you see, you're well on your way to selecting a science fair project about insects.

## Using this Book

The next thing to do is to look through the "Table of Contents" in this book for topics to research. This book contains 20 science fair projects involving insects. Read through the "Background" and "Project Overview" sections of each project. Every project in this book can be adjusted, expanded upon, or fine-tuned in some way in order to personalize your investigation. After reading through these sections, think about how you can put your own signature on the experiments. The "Going Further" and "Suggested Research" sections of this book are designed to help you personalize each project.

**Fig. 3-1** *Insect behavior can be observed almost anywhere such as this ant colony in pavement cracks.*

## Other Sources

At this point you should be narrowing down the topic that you want to research. You can begin your project now or continue to look for more insight into the problem you want to study. Consider branching out even further by looking through science sections of newspapers, such as the science section of the Tuesday New York Times. Also look at magazines such as *Popular Science*, *Discover* or *Omni*, which cover a broad range of topics, or *E Magazine*, *Buzzworm*, *Garbage*, or *BioCycle* which concentrate on interdisciplinary environmental topics. Check the *Reader's Guide to Periodical Literature*, which is an index that lists articles in numerous magazines and gives a brief synopsis of these articles. Your school textbooks might also be helpful. Check references to other books, usually found at the end of each chapter.

Other helpful sources include educational television shows such as "NOVA," National Geographic specials, "Nature," and many others. Almost all of these types of shows are found on public television and cable networks. Check your local listings to see what might be viewed in the near future in your area. Also, don't hesitate to use past science fair projects as a source of interesting topics (the previous chapter on scientific research covers this source in more detail).

## Talking to Specialists in the Field

Once you have a good idea for a project, consider talking with a professional. For example, if your project involves pesticides, arrange a meeting with an agricultural specialist who works for your state or county, a professor of entomology at a nearby university, and a farmer who uses pesticides. If your project involves oil spills, speak with a professional who studies the problem, people affected by the problem, and those who caused the problem. If you are studying recycling, speak with people who recycle and those who don't, and with scientists who study the process. Interesting science fair projects involve not only equipment, chemicals, and cultures, but also what people think about the topic—pro, con or indifferent.

Be sure to use any resources that are readily available. If you live near a sewage treatment plant, landfill, agricultural research station, large mechanized farm, small organic farm, or any other type of facility that can contribute to your project, try to use it to your advantage. (See FIG. 3-2.). If you have a parent or friend who is involved in a business or profession applicable to your project, try to incorporate it into your research. The most important thing to remember is to select a project that you are truly interested in learning more about.

**Fig. 3-2** *Any resource, such as farms and power plants, can be put to use in a science fair project.*

## Putting Your Signature on the Project

All the projects included in this book are good candidates for a science fair. What could make these projects outstanding examples of research is the way in which you put your signature on them. What ideas for further research have you found in the "Going Further" section, or did you delve into the "Suggested Research" section? Did a teacher, scientist, or businessperson add an interesting aspect of the research to make it truly unique and your own?

## Before You Begin

Review the entire project with your sponsor in order to anticipate problems that might arise. Some projects must be done at a certain time of year. Some can be done in a day or two, whereas others can take a few weeks, months, or even longer.

Some projects use supplies that are found around the home, but many require equipment or supplies that must be purchased from a local hardware store, science/nature store, or a scientific supply house. Some projects require organisms such as microbial cultures or insects. Your sponsor might have access to the organisms needed for the project. Organisms like insects might be caught in the wild, bought at a pet or bait shop, or ordered from a scientific supply house. Bacterial cultures might be available from your school or could be ordered from a supply house.

Each project in this book states not only what organisms and equipment are needed but also how or where they might be procured. Much of the equipment and supplies—such as collecting nets, petri dishes, magnifying glasses, and microscopes—might be available at your school and need not be procured on your own. Some projects require the rearing or culturing of organisms and demand considerable attention before your research begins.

Plan ahead financially too. Look through the "Materials" section of each experiment. Be sure to include any materials needed for additional research that you have added to the original project. Determine how and where you will get everything and how much it will cost. If a dissecting microscope is needed, do you have access to one? If a live insect is needed, can you catch it in your location during that time of year or must it be ordered from a supply house? If you need a bacterial culture, is it available from your teacher or a nearby university or must you purchase it? How much will these items cost? Don't begin a project unless you can budget the appropriate amount of time and money as suggested by your sponsor.

## Getting Started

Once you have selected a project by following the suggestions in the previous section, use the following suggestions to help you get organized.

## Scheduling

Before proceeding, it is a good idea to develop a schedule to help ensure that you have a completed project in time for the fair. Have your sponsor approve your timetable. Leave yourself time to acquire the needed equipment, supplies, and organisms. A generalized timetable is given below.

Most science fair projects require at least a few months of preparation to complete. In many instances they can (and must) be completed in less time. It would be difficult to produce a prize-winning project, however, without plenty of time.

- Identify your adult sponsor.
- Choose a general topic.
- Use a notebook for all note taking throughout the project.
- List resources (including libraries to go to, people to speak with, businesses, organizations, or agencies to contact, etc.).
- Select reading materials, use bibliographies for more resources, and begin a formal literature search.
- Select the exact project and develop a hypothesis, write a detailed research plan and discuss it with your adult sponsor; have your sponsor sign off on your final research plan.
- Procure equipment, supplies, organisms, and all other materials.
- Follow up on your resource list: speak with experts, make all contacts, etc.
- Set up and begin experimentation.
- Begin to plan for your exhibition display.
- Collect data and rigorously take notes.
- Begin writing your report.
- Begin to analyze data and to draw conclusions.
- Complete your report and have your sponsor review it.
- Design your exhibit display.
- Write your final report and abstract and be sure your notebook is available and readable.
- Complete and construct a dry run of your exhibit display.
- Prepare for questions about your project.
- Disassemble and pack your project for transport to science fair.
- When the fair starts, set up your display and enjoy yourself.

## Literature search

As you can see from the suggested schedule, one of the first items is to perform a full-blown literature search of the problem that you intend to study. A literature search (also often called research) means reading everything that you can get your hands on about a topic. (See FIG. 3-3.) Read newspapers, magazines, books, abstracts, and anything related to the specific topic to be studied. Use on-line databases if they are available. Talk to as many people as possible who have some insight into the topic. Listen to the news on radio and television. At this point you might want to narrow down or even change the exact problem that you want to study.

**Fig. 3-3** *The most valuable tool at your disposal is your school or local library.*

Once your literature search is complete and you have organized the data both on paper and in your mind, you should know exactly what problem you intend to study and should formulate your hypothesis.

## The research plan

At this point you should have completed a research plan. You can use portions of this book to get started with your research plan, but you must go into additional detail and include all modifications. Before beginning the project, go through it in detail with your adult sponsor make certain that the requirements of the project are safe, attainable, suitable, and practical. In many science fairs, your sponsor is required to sign off on the research plan, thus attesting to the fact that it has been reviewed and approved.

It is important to review your particular fair regulations and guidelines so that your project won't run into any problems as you proceed.

## Science fair guidelines

Almost all science fairs have formal guidelines or rules. Check with your sponsor to see what they are. For example, there may be a limit on the amount of money that can be spent on a project or a moratorium on the use of live (vertebrate) animals. Review these guidelines and make sure that the experiment poses no conflicts.

Many science fairs require four basic components for all entries: (1) the actual notebook used throughout the project that contains data collection notes which may be read by fair officials; (2) an abstract of the project (usually no more than 250 words long) that briefly states the problem, proposed hypothesis, generalized procedures, data collection methods, and conclusions; (3) a full-length research paper (sometimes not required); and (4) an exhibition display.

## The research paper

A research paper might be required at your fair, but consider doing one even if it isn't necessary (you might be able to get extra credit for the paper in one of your science classes). The research paper should include seven sections: (1) a title page, (2) table of contents (3) an introduction, (4) a thorough section on procedures, (5) a comprehensive section discussing what you did as well as what you thought while doing the research and experimentation, (6) a separate conclusion section that summarizes your results, and (7) a reference and credit section in which you list your sources and give credit to anyone who helped you or to any company, organization, or agency that assisted you. There are many good books listed in the sources section of this book that detail how to write a research report.

## The exhibition display

The exhibition display should be as informative as possible. Keep in mind that most people, including the judges, will only spend a short while looking at each presentation. Therefore, try to create a display that gets as much information across with the least amount of words. Use graphs, charts, or tables to illustrate data. As the old saying goes, "A picture is worth a thousand words." Make the display as attractive as possible since you cannot communicate the value of your project if you don't draw people's attention to it (see FIG. 3-4).

Discuss with your sponsor such exhibit requirements as special equipment, electrical outlets, and wiring needs. Live organisms of any kind are usually prohibited from being displayed. Often, preserved specimens are also prohibited. Usually no foods, wastes, or even water is allowed in an exhibit. No flames, gases, or harmful chemicals are allowed either. Before proceeding find out what you can and cannot do.

Many fairs have specific size requirements for the actual display and its backboard. For more information on building an exhibition display, see the reference section at the end of this book. (See FIG. 3-4.)

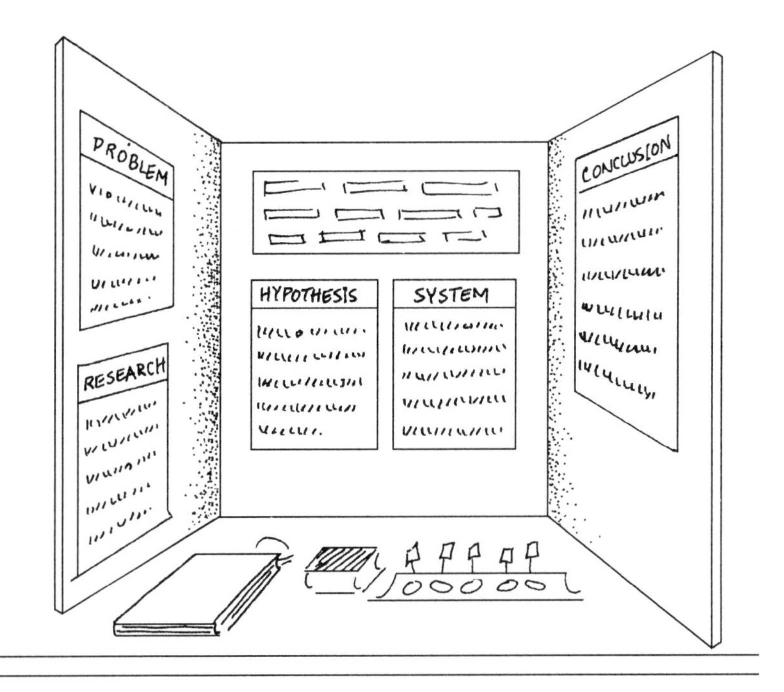


**Fig. 3-4** *The science fair display is the culmination of all your work.*

## Judging

Adherence to the scientific method and attention to detail is crucial to the success of any project. Judges usually want to see a well-thought-out project and a knowledgeable person who understands all aspects of his or her project.

Most science fairs assign a point value to various aspects of a project. For example, the research paper might be worth "X" points while the actual display might be worth "Y" points. Request any information that might give you insight about the judgment criteria at your fair.

## PART II

# Insect Lives

There is no better way to begin thinking about science fair projects involving insects than to look at a broad spectrum of topics—which is what the first part of this book tried to do. You can think of these first four projects as a broad look at insect lives and the insect success story that has been going on for the past 400 million years. Not bad if you compare that with human beings who have only been around about 1 million years.

Looking through these first few projects should begin to give you a feel for the diversity of topics available when studying insects. One project studies growth hormones in insects. It investigates the insect's anatomy and physiology and lets you actually manipulate and control the life cycle of an insect. Another project investigates which foods are used most efficiently by an organism. Using insects makes the experiment easy to do, but the ramifications are far-reaching and can be related to almost any other type of animal.

One of the most amazing aspects of insects is their fecundity, or their ability to produce offspring. A theoretical "doomsday chart" illustrates this point dramatically in still another project in this section. Finally, a more down to earth project determines honeybee color preferences and honeybees' ability to detect sugar concentrations. Why should a plant waste its energy supply if the bees can't tell the difference?

# 4

# Hormones

## Growing up is hard to do

*(What is the source of the blow fly growth hormone and when is it produced?)*

Hormones are used by all animals, including insects, for functions such as controlling the growth and timing of reproduction. Insect hormones are released into the *hemolymph* (insect blood), where they are carried throughout the insect's body and transported to the organs. Before studying insect hormones, you must understand two facts about insect physiology. First, insects have an open circulatory system, which means that the blood isn't carried by tubes throughout the body as with higher forms of life. Instead, the blood simply bathes all the body tissues as it flows over them.

Second, an insect's blood doesn't carry oxygen. Oxygen is transported to cells via a series of tubes called the *tracheal system.* Since hemolymph does not carry oxygen, an insect can survive a while with its circulation restricted during a process called *ligation.* Ligation allows you to separate various parts of an insect body, restricting blood flow, yet still allowing the insect to survive.

## Project Overview

All insects produce a molting hormone (called *ecdysone*) that controls when an immature insect molts (sheds its exoskeleton) during the process of metamorphosis. A second hormone called "juvenile hormone" controls what the insect will molt into—another larva or a pupa. When juvenile hormone is present, the larva molts into a larger larva. When the hormone is absent, the larva molts into a pupa.

Using the ligation technique described below, you can study hormone flow through the insect's body. Since the ligation restricts the flow of hemolymph, it also restricts the flow of any hormone carried in the hemolymph, thus preventing the horomone from reaching certain organs.

In this experiment, you will first ligate blow fly larvae at various locations. Portions of a larva body that don't receive the molting hormone (due to the ligation) won't be able to molt. By using this technique, you'll be able to locate where the hormone is being produced. Once this is done, you will determine how far in advance of pupation the hormone is actually released.

Where is the molting hormone produced in the blow fly larvae? How far in advance of pupation is this hormone released? To answer these and any other questions that your research leads you to, begin your literature search and formulate your hypothesis.

## Materials

- 10 to 20 blow fly larvae (*Phormia*) that are at least three or four days away from pupating. (You only need about 10, but might need extras until you get your ligation technique working properly. The flies can be ordered from a scientific supply house.)
- A spool of strong, fine thread
- Scissors
- Five petri dishes
- Water
- Paper toweling
- Scalpel (for the "Going Further" section, only)

## Procedures

This experiment can be done at any time of year, but it should be performed immediately after receiving the fly larvae. In the first part of the project, you will ligate fly larvae at various locations along the body in order to determine which part of the body is producing the molting hormone. In the second part, you will determine when the hormone is released from that part of the body that you've identified in the first part of the project.

For the first part, cut five pieces of thread, each about 6 inches long. You will take four blow fly larvae and ligate them at different locations using the technique described below. You will ligate one larva just behind the head, another a quarter of the way back, another midway, and the last one, three quarters of the way toward the rear. The object is to narrow down the location of the organ that produces the molting hormone.

To ligate each larva, place it in a petri dish half-filled with water. (The water slows down its movement.) Place the thread around the insect's body (at the proper location) and tie the thread into a tight knot so that the insect's body is almost completely constricted. Make sure however, that the thread does not cut into the cuticle (skin). (See FIG. 4-1.) This constriction prevents the hemolymph from passing through the ligated area. Place some damp paper toweling on the bottom of a petri dish (it can be the same dish) and place the ligated fly larvae in the dish. (See FIG. 4-2.)

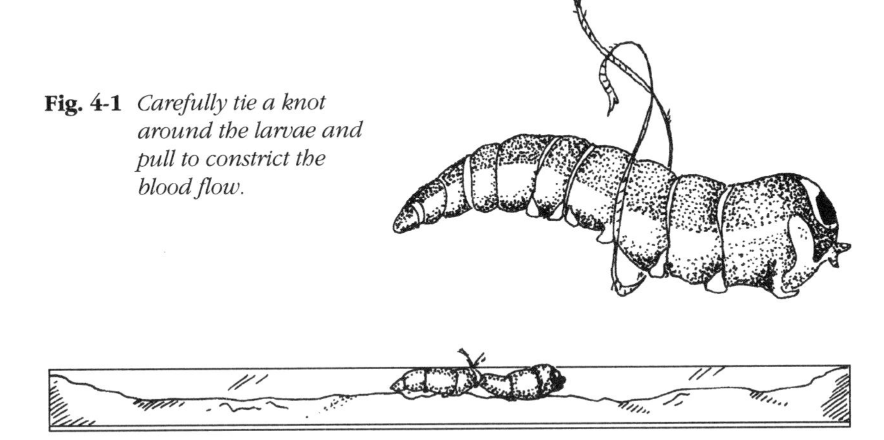

**Fig. 4-1** *Carefully tie a knot around the larvae and pull to constrict the blood flow.*

**Fig. 4-2** *Place the ligated fly into the bottom of a petri dish that has a layer of damp paper towels on the bottom.*

Perform the same procedure for each of the four ligation locations as mentioned above, so that you have four ligated flies in four petri dishes. Place a few unligated larvae in a fifth petri dish as a control. Check the larvae every day. When they begin to pupate (they'll turn brown), observe what has happened with the experimental larvae. (See FIG. 4-3.) Observe the form and color of all the flies.

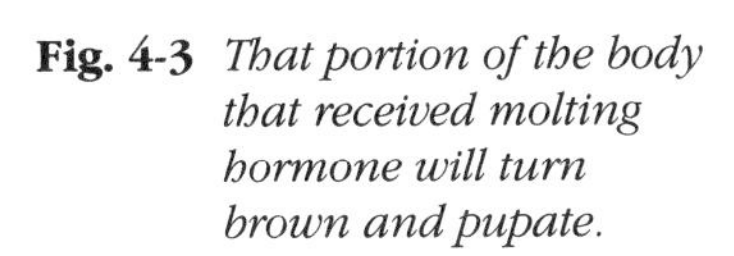

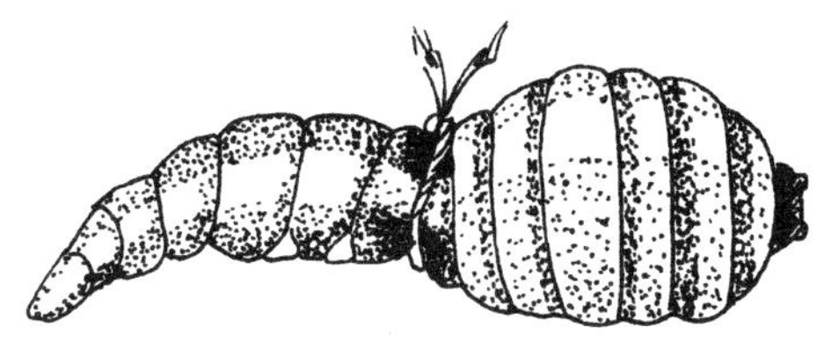

**Fig. 4-3** *That portion of the body that received molting hormone will turn brown and pupate.*

Once you have identified the region of the body that releases the molting hormone, determine how many hours before molting the hormone is released into the blood. To do this, ligate (midway down the body) larvae at intervals of 24 (or 12) hours prior to pupation. (Begin at least four days prior to pupation or as soon as the fly larvae arrive in the mail.) Use the ligation procedures described above and place each ligated larva in a petri dish as described above. Mark each dish with the date and time that the ligation was performed. Which larvae molted into a pupa (pupated) and which did not molt at all? At what point did ligation no longer prevent pupation from occurring? In other words, the hormone was already released and flowing throughout the insect's body when the ligation was performed.

## Analysis

What happened to the larvae ligated at different places along the body in the first part of the project? Which portions of the body pupated and which did not? By deduction, what part of the body must be producing the molting hormone, and by your research, which organ? From the second part of the experiment, how far in advance of pupation is the molting hormone released into the hemolymph?

## Going Further

What would happen if a larva had both juvenile and molting hormones artificially introduced into the body? When would it molt and what would it molt into—a pupa? Or would it continue to molt into a larger larva? This can be tested by taking a larva that is about one day away from pupating and another younger larva that would normally molt into another larva, and transferring hemolymph between the two.

To do this, make a small incision along the side of the larva about to pupate and a similar incision along the side of a small (younger) larva, which would not normally pupate at this time. Hold both incisions up to each other and gently squeeze the older larva so hemolymph is forced out and into the younger larva's incision.(See FIG. 4-4.) Place the younger larva in a petri dish as described above and observe what happens over the next few days.

## Suggested Research

- Research how various growth hormones are used for pest control.
- Research integrated pest management and biological control of insect pests.

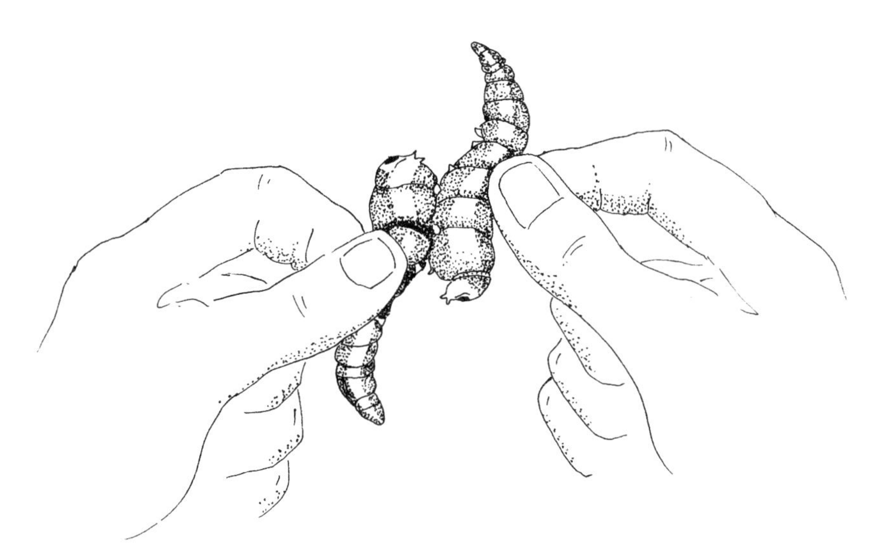

**Fig. 4-4** *Hold the two larvae together at the incisions and gently squeeze the older larva to inject fluid into the younger larva.*

# 5

# Nutritional value of foods

## You are what you eat

*(Comparison of nutritional value of an artificial diet and a natural diet when eaten by tomato hornworms)*

Have you ever wondered why poultry is usually quite inexpensive? Part of the reason is because a 5-pound chicken can be raised from birth on less than 10 pounds of feed. This is due, in part, to the nutritional value of the feed, which is readily converted into the chicken's own tissues. Artificial diets are used for rearing many domesticated animals, artificial nutrients are used for growing plants hydroponically, and artificial foods are used for sustaining astronauts in space.

Some foods are more digestible and are more efficiently converted into an animal's tissues than others. When creating artificial diets—or when simply trying to determine the best food source for an animal—it is helpful to have some method for comparing the quality of foods. There are two commonly used methods that measure the suitability of a food source for a particular organism. The Efficiency of Conversion of Ingested Foods (ECI) measures the quality of the food—that is how much of the food was converted into the animal that ate it. The other method is called Approximate Digestibility (AD) and measures how easily a food can be digested.

### Project Overview

In this project you will compare the artificial and the natural diet of the *tomato hornworm* in order to understand how ECI and AD measurements are taken. The ECI compares an organism's weight gain to the weight of the food that it has eaten. The AD takes into consideration how much of the food was lost as waste (the more waste, the less digestible the food).

The ECI and AD indicators can be used to study the value of many types of artificial foods and to compare them with an organism's natural foods. This project will determine whether an artificial diet is a more digestible and efficient nutritional source for a tomato hornworm than a natural food diet. (See FIG. 5-1.)

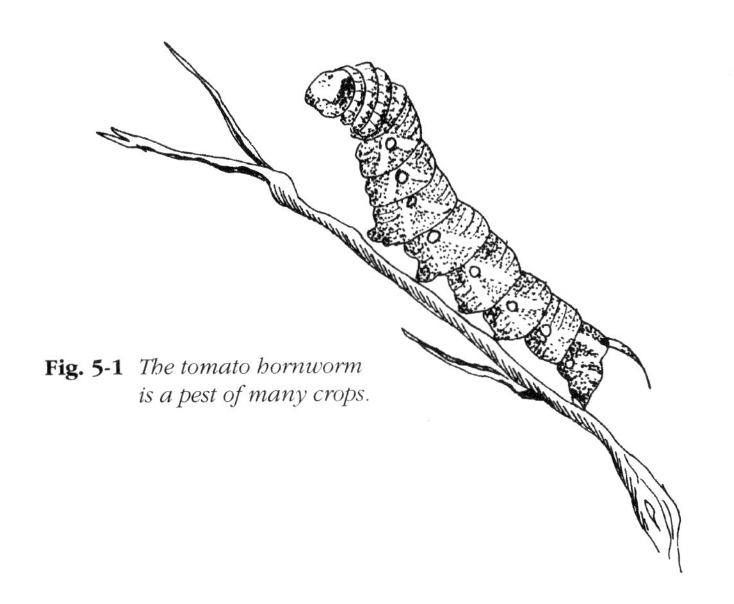

**Fig. 5-1** *The tomato hornworm is a pest of many crops.*

Is an artificial diet for a tomato hornworm as digestible and efficiently processed as the insect's natural foods? To answer these and any other questions that your research leads you to, begin your literature search and formulate your hypotheses.

## Materials

- 10 tomato-hornworm (also called tobacco hornworm) larvae at least 1 inch long, but preferably 2 inches long (available from a scientific supply house)
- Artificial hornworm diet (also available from a supply house)
- Five potted tomato plants about 7 inches tall
- An electronic scale
- 10 translucent plastic cups (16 oz.)
- Flexible screen or mesh to cover the cups
- Rubber bands to hold the mesh on the cups
- A few desk lamps

## Procedures

This project can be done any time of year if you plan to purchase the hornworms and have access to potted tomato plants. When you obtain the hornworms, raise them on the artificial diet until they are 2 inches long (if they are already 2 inches long, skip this step). Once they reach the required length, remove them from their existing containers and put each in an empty plastic cup. Cover the cups with mesh and hold in place with a rubber band. (See FIG. 5-2.)

**Fig. 5-2** *Each larvae should be placed in its own cup with food and covered with nylon mesh.*

After 24 hours, remove the larvae from each cup and clean out any feces that accumulated in the cups. Now, place 10 grams of artificial diet in each of 5 cups. In 5 more cups, place 10 grams of the hornworm's natural food, tomato plant leaves (you can tear up the leaves to get them to fit into the cup).

Place one hornworm in each of the 10 cups. Be sure to number each cup with the type of food in it. If you place a lamp above each group of cups, the insects will probably eat more, thus producing better results. Be sure that the lighting is the same for the two groups of cups.

Wait another 24 hours and then begin collecting the following data from each cup: (a) the weight of the food that was placed in the cup (10 grams), (b) the weight of the food remaining in the cup, (c) the weight of the food eaten by the hornworm (a – b = c), (d) the weight of the feces in the cup, and (e) the weight of the insect. Use TABLE 5-1 for entering your data.

When you have documented all these figures for each cup, discard everything in the cups, except for the insect, and repeat the procedure.

**Table 5-1**

| INSECT #1 - TOMATO PLANT | | | | | | |
|---|---|---|---|---|---|---|
| Day | Wgt. of food given | Wgt. of food remaining | Wgt. eaten | Wgt. of feces | Wgt. of insect | |
| 3 | 10 gr. | | | | | ← Start wgt. |
| 4 | 10 gr. | | | | | |
| 5 | 10 gr. | | | | | |
| 6 | 10 gr. | | | | | |
| 7 | 10 gr. | | | | | ← End wgt. |
| | | | | | | ← Total wgt. gain |
| | | | ↑ Total wgt. eaten | ↑ Total wgt. feces | | |

Be sure to keep the insects in the same cups, feeding on either the artificial or natural diet. Each day, take the same measurements and repeat the same procedure. Do this for 5 more days, after which time you can analyze the data.

## Analysis

Determine the overall quality of the food by finding out how much of the food was converted into hornworms. To calculate the ECI for each hornworm, use the following formula: total weight gained over 5 days/total weight of food eaten over 5 days. Multiply the result by 100. The more food was actually used to make the insect grow, the closer the EC1 is to 1.

Determine how easily digested the foods were by calculating the AD for each hornworm, using this formula: (total weight of food eaten − weight of feces)/total weight of food eaten. Multiply the result by 100. The closer the AD is to 1, the more digestible the food.

Once the ECI and the AD have been calculated, average together the numbers from the 5 larvae eating the artificial diet and average together the numbers from the 5 larvae eating the natural food.

How did the two groups compare? Which type of food is more efficiently absorbed and digestible? How great are the differences?

## Going Further

- Now that you know how to determine the value of a food for a particular organism, design another experiment using other organisms and other foods.

- Perform the same experiment as stated in this project, but concentrate on other factors, such as the health of the insects and their ability to reproduce. Does an artificial diet have any side effects?

## Suggested Research

- Study how it is possible for a chicken to convert half of everything it eats into its own body mass?
- Research what other factors besides artificial diets are used to increase the growth rate of domesticated animals, such as cows.

# 6

# Population growth
## A doomsday chart

*(Population growth of aphids)*

Studying the population growth of a species helps us to understand how that species fits into our biosphere. What environmental factors cause illness (morbidity) and death (mortality) in a species, thus controlling the size of its population? Entomologists and ecologists must study the population growth of insects in order to know how they survive, how they behave with other organisms, and how they fit into their ecosystem.

Population growth left unchecked results in what is called a J growth curve, since the population starts small (the lower portion of the J) and then continues an upward trend. (See FIG. 6-1.) If organisms continued this unlimited growth, they could theoretically take over the world. To dramatically illustrate the J-curve trend, a "doomsday chart" could be created for organisms, including many as small as insects.

## Project Overview

Many insects have extraordinary powers of reproduction—which is one of the reasons that they are so successful and numerous. They can produce incredible numbers of offspring in short periods of time. This reproductive capacity allows them to recover from most disasters very rapidly.

*Aphids*, also called plant lice, have some of the most remarkable reproductive capabilities. (See FIG. 6-2.) During most of their life cycle, they reproduce without mating (called *parthenogenesis*) and are born already pregnant (called *paedeogenesis*).

For most of the year, an aphid population is entirely composed of females and produces young by parthenogenesis. Instead of laying eggs,

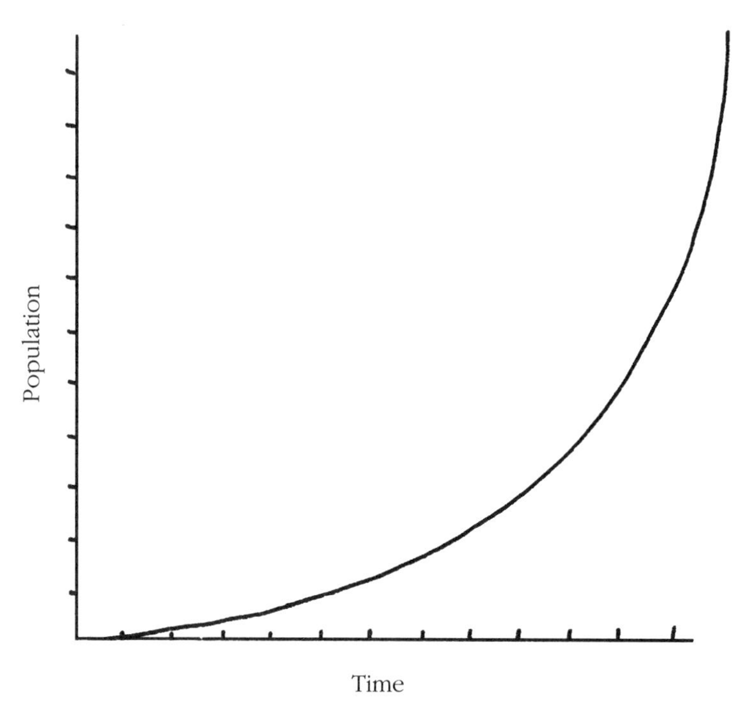


**Fig. 6-1** *A J-curve of population growth represents a population explosion left unchecked by natural forces.*

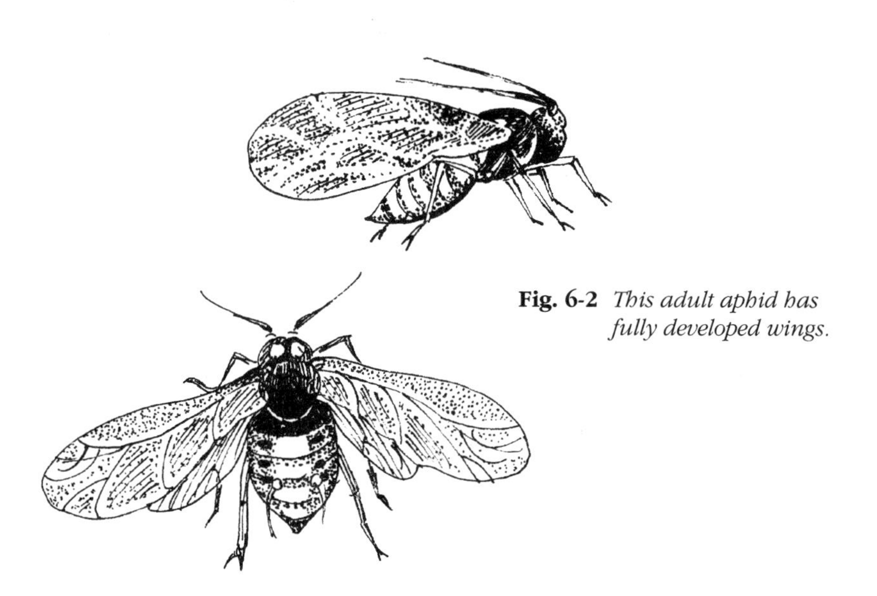

**Fig. 6-2** *This adult aphid has fully developed wings.*

like most insects, the female aphid produces live young that are already pregnant with the next generation of offspring. The young aphids begin to feed immediately, and after molting into adults, they too are giving birth.

In this project, you will first perform an experiment to see how many offspring a single female aphid can produce in her lifetime. Then, in the second part of the experiment, you will mathematically calculate a "doomsday chart" using the data that you collected in the first part of the experiment. This chart will determine how long it takes for a single female aphid to generate a population of one million insects and beyond, assuming no mortality.

How many offspring can a single aphid produce? How long would it take for one aphid to produce a population of one million? How much would these aphids weigh? To answer these and any other questions your research leads you to, begin your literature search and formulate your hypotheses.

## Materials

- A plant infested with aphids (found in fields and gardens, or produced by placing a bean plant outdoors in a very sunny spot during the summer months).
- Nylon material such as pantyhose
- String
- Forceps or fine tweezers
- Small paintbrush (the kind used to paint model planes)
- Magnifying glass
- Calculator
- Graph paper
- Electronic scale (optional)

## Procedures

The first part of this experiment involves using actual insects and must be done from late spring to early fall (you might be able to find aphids in a local greenhouse during the colder months). The second part is theoretical and can be done at any time of the year. You will select, isolate, and observe one aphid in a colony for 2 to 4 weeks.

To begin, on the stem containing the colony, carefully look for a small individual, which will be an immature form. It should not have wings. (See FIG. 6-3.) You might need a magnifying glass to see whether the aphid has wings. To isolate this individual, use the brush or tweezers to push off all the other aphids nearby. You want this individual to be clear of other aphids in the colony by at least 3 inches in both directions on the stem.

**Fig. 6-3** *This immature aphid has no wings.*

If you pick up the selected aphid while its beak is inserted into the plant, you will tear off its mouthparts and kill it. If necessary, using the brush you can gently pick up an aphid and place it on a stem where there are no other aphids.

When an individual is isolated, with no neighbors within 3 inches of it, enclose that part of the plant stem (with the single aphid) in a fine nylon material, such as pantyhose, so that no other aphids can crawl or fly into that area. (See FIG. 6-4.) To do this, cut a nylon strip, 4 inches wide, and wrap it around the stem, so that it surrounds the plant loosely. Don't bind the plant; be sure to leave some room for the aphid. Use string to tie each end of the nylon net to the plant stem. Make it tight enough so that aphids can't crawl in or out, but don't crush or break the stem.

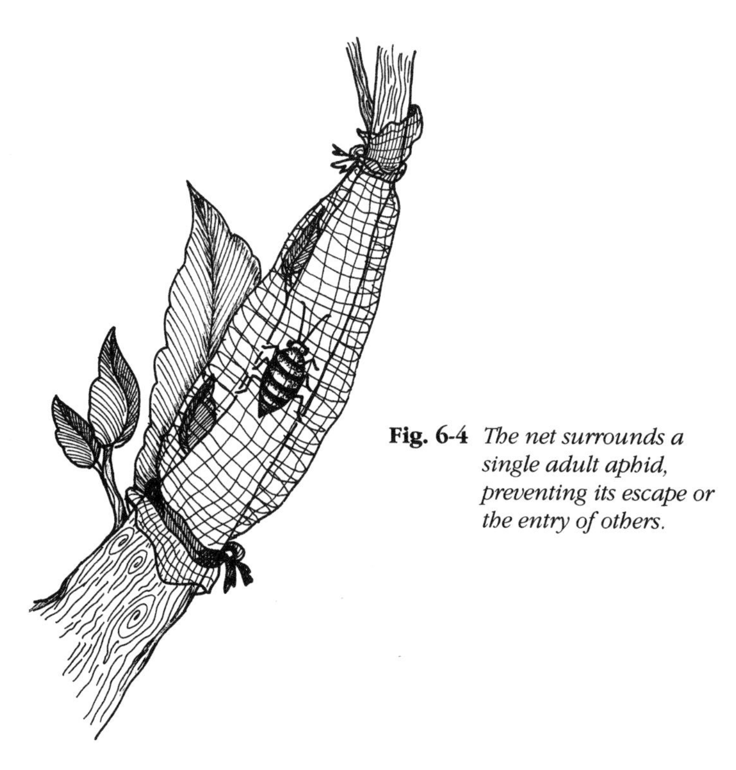

**Fig. 6-4** *The net surrounds a single adult aphid, preventing its escape or the entry of others.*

Create a similar setup on a different plant, or at least a different part of the same plant, so that you'll have two experiments running concurrently in case a problem occurs with one. Be sure to write down the date and time that you begin the experiment. Return every 2 to 3 days and count any new aphids (offspring) in the enclosures. To do this, gently untie and remove the nylon so that you can get a clear view of the stem and the aphids. Count the new offspring and then remove them from the enclosure with the brush (the offspring will be smaller than the mother). Once again try not to crush them as they are removed. If the original aphid falls off the stem, use the brush to carefully place it back on the stem. Replace the netting over the stem after each observation. Do this procedure with each of the plants being tested.

Continue your observations until offspring are no longer being produced. (The original aphid will die shortly thereafter.) Keep a record of the date and the number of offspring for each aphid. How many offspring did each aphid produce? How long did it take to produce the offspring?

With the information gleaned in the first part of the project, you can then perform the second part of the project. Calculate how long it would take to get one million aphids starting with a single female and assuming all her offspring survived. Graph this J-curve.

Finally, overlay your population graph with a biomass chart. How much would the population of aphids weigh when it reached one million individuals? Weigh about 50 individuals on an electronic scale and then divide to get each individual's weight. If you cannot weigh your specific individuals, assume each to weigh .00082 grams. What would the number and weight of the aphid population be after a year or more?

## Analysis

How many offspring did a single aphid produce? How many offspring would theoretically be produced in future generations if they all survived? The results should represent a typical J-curve. How long would it take for the population to grow to one million individuals? How about 10 million or one billion? When you have these numbers, determine what the weight of these insects would be and how their weight would compare to the weight of an ecosystem such as a deciduous forest, one mile square.

With few exceptions, morbidity and mortality factors keep populations from actually creating a J-curve. Environmental resistance such as predators, disease, and toxic substances result in a population growth called an S-curve. What types of organisms, if any, have populations that truly produce a J-curve?

## Going Further

- After you calculate the numbers under ideal conditions, be more realistic. Vast numbers of insects don't survive to reproduce. Some are eaten by predators, some die of disease, and some will succumb to the environment. Incorporate into your figures assumptions about mortality. For example, assume that 30% of all the young are destroyed by predators, disease, harsh weather conditions, etc. How does this affect the curve?

- Devise an experiment that incorporates mortality factors into the growth curve.

## Suggested Research

- Read about the growth curve of Homo sapiens.
- Read about the study of population dynamics.

# 7

# Honeybees
## In search of food

*(Honeybee attraction to color and sugar concentrations)*

Pollination by insects is crucial to most natural ecosystems and many cultivated plants. Although many different types of insects pollinate plants, *honeybees* are the most well known and important. We depend on honeybees to pollinate fruit trees, small berries, and many forage plants that our domesticated animals eat. These plants cannot produce seeds or fruits without first being pollinated by a bee. Plants get insects to pollinate them by producing nectar—a sugar solution that the insects use for food. Nectar is the reward that plants give to bees for their pollination services. Bees fly from flower to flower collecting the nectar. As they do this, they pick up pollen from one flower and deliver it to another flower of the same species. This transfer of pollen results in seed production and plant reproduction.

## Project Overview

This project is designed to first determine what colors attract honeybees and to then see if they can differentiate between various concentrations of sugar when they have found the source.

Since a plant must use its own sugar supply to produce nectar, the more sugar for the nectar the less food for the plant. Thus, a high sugar concentration in the nectar is costly to the plant. It would therefore be beneficial for a plant to produce nectar with as little sugar in it as possible, yet still attract the bees.

What colors attract honeybees to flowers? When honeybees arrive at a flower, can they differentiate different sugar concentrations? Is there a minimum threshold concentration that the bees detect? To answer these and any other questions that your research leads you to, begin your literature search and formulate your hypotheses.

CAUTION: This experiment involves observing bees. Follow the instructions carefully and consult with your sponsor be fore beginning. If you are allergic to bee stings, do not at tempt this experiment.

## Materials

- Access to an area that is often inhabited by honeybees (Discuss this aspect of the project with your sponsor and follow safety precautions as suggested by your sponsor while performing this experiment.)
- One piece of posterboard in each of the following colors: yellow, red, blue, green, and black
- Three pieces of white posterboard
- Glue
- Tape
- Eyedropper
- Outdoor table
- Scissors
- Six glass saucers (or the bottoms of petri dishes) about 4 inches in diameter, that can hold about ½ inch of fluid
- Six containers for mixing sugar solutions (capable of holding about 2 cups of liquid each)
- Granulated sugar
- Teaspoon and tablespoon
- Measuring cup
- Heavy garden gloves
- Binoculars (optional) (to observe the bees from afar)

## Procedures

This experiment can be done from late spring through the summer. In the first part, you'll do a test to determine whether honeybees are attracted to one color more than others. When you know the honeybees' color preferences for locating their food, you'll do a test to determine whether they have preferences about sugar concentrations.

To begin the first part of the project, use the bottom of a petri dish to draw a circle on each of the colored pieces of posterboard. Cut the circles out and glue each one to a white piece of posterboard. Each circle should be evenly distributed and separated by a few inches.

Next, on each of the circles you'll place dishes containing a sugar-water solution. In a bowl, mix 4 cups of water and 12 teaspoons of granulated sugar. Stir the solution thoroughly and pour about ⅛ inch of the liquid into each petri dish (or saucer). You'll need one dish for each colored circle.

On a warm sunny day, find a location where honeybees are present—a sunny open meadow or lawn will do well. Set up the table, place the posterboard containing the colored circles on it, and tape the board down so that it won't blow away. Place a dish containing the sugar solution on each circle and move a certain distance away from the table when you make your observations. (See FIG. 7-1.) You can stand a few yards away and watch with the naked eye, or you can stand farther away and observe with binoculars.

When the bees begin to arrive, observe them for at least 30 to 45 minutes. Write down the starting time of the experiment and the approximate number of bees visiting each dish. It is important to distinguish which dish each bee visits first. Also, watch to see if some dishes run out of liquid before others.

**Fig. 7-1** *Place the petri dishes on the colored circles of the posterboard and observe the bee activity from afar.*

Write down the time when a dish empties. Prioritize the colors selected by the bees. When you have determined what colors attract the bees to the sugar, begin the second part of the experiment to determine whether the concentration of sugar is important to the bees.

Make five different sugar solutions. In each bowl, mix 2 cups of water with the following amounts of sugar: 1½ teaspoons (concentration #1), 3 teaspoons (concentration #2), 6 teaspoons (concentration #3), ¼ cup + 3 teaspoons (concentration #4), and ½ cup + 6 teaspoons (concentration #5). Stir the solutions thoroughly until all the sugar is dissolved (you might need to heat the solution). Label 5 petri dishes with the proper concentration number. Finally, label a sixth dish "plain water." These mixtures should be ready before you go on location to perform the experiment.

Now pour 2 tablespoons of each solution into the properly marked petri dish. (See FIG. 7-2.) Be sure to clean the spoon between each use. Next, from the posterboard cut out six circles of the color preferred by the bees in the first part of the experiment. Attach the circles to a white posterboard in the same fashion as in the first part of the project. These circles will be placed beneath the dishes containing the various sugar concentrations. Attach the new posterboard (containing six circles of the same color) to the table. Place each dish containing two tablespoons of solution on each circle and be sure that each dish is labeled with the correct concentration number. These numbers must be visible during your observations. When all is ready, begin your observations but remain at a safe distance from the table.

When bees arrive, note the number of bees near or in each dish. In this part of the experiment, it is more important to note how many insects are in or around any given dish, not which dish they arrived atfirst. Every 5 minutes, write down the number of bees at each dish. Continue your observations for one hour. Note the time when any of the dishes become empty. Record your data in TABLE 7-1.

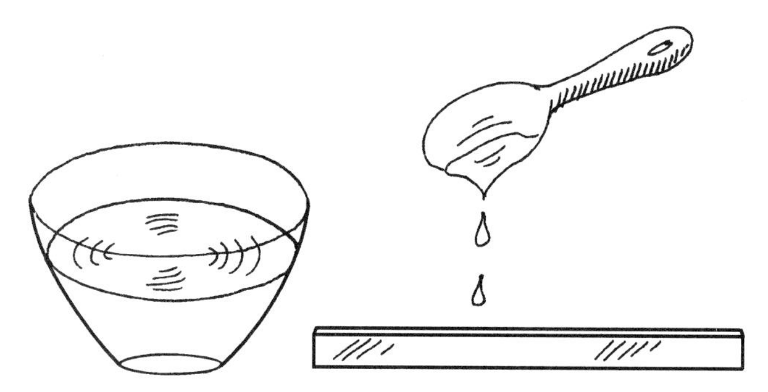

**Fig. 7-2** *After mixing the proper solutions, pour two tablespoons into the correctly marked petri dish.*

CAUTION: Be very careful when removing the dishes. Use gloves when picking up the dishes because bees may be on or in them.

**Table 7-1**

| | TIME IN MINUTES | | | | | | | | | | | |
|---|---|---|---|---|---|---|---|---|---|---|---|---|
| Concentration | 5 | 10 | 15 | 20 | 25 | 30 | 35 | 40 | 45 | 50 | 55 | 60 |
| 1 | I | | | | | | | | | | | |
| 2 | III | | | | | | | | | | | |
| 3 | Ø | | | | | | | | | | | |
| 4 | ~~IIII~~I | | | | | | | | | | | |
| 5 | II | | | | | | | | | | | |
| 6 | III | | | | | | | | | | | |

## Analysis

Did one color seem the most attractive to the honeybees? Was there more than one color that they preferred? When the bees landed on a dish, could they detect different concentrations of sugar? Did the bees visit all the dishes equally? If not, which concentration was visited the most and which the least? Analyze your data.

From this information, can you conclude that bees could sense the different concentrations of sugar? Is there a minimum sugar concentration (threshold) that they require? Is there a maximum concentration? How might a maximum concentration be important to the plant?

## Going Further

- Continue this experiment by placing posterboard of the bees' favorite color under the sugar concentration that the bees visited most. For a few days, refill this dish with the sugar water at the same time of day. Then, move the colored paper from beneath the sugar-water dish to beneath a dish containing plain water. Leave the sugar-water dish on a piece of white posterboard. Which dish do bees visit first: the plain water dish over their favorite color or the sugar-water dish over no color? What do you think the results indicate?
- Continue this experiment by using different shades of the color found to be the most attractive to the bees. You can lighten the color by using paint instead of posterboard. Add varying amounts of white paint to the original color to obtain various shades. Can insects distinguish between certain shades? Do they find a particular shade of their fa-

vorite color more alluring? Do flowers appear to take advantage of bees' color preferences?

## Suggested Research

- Read about insect vision.
- Investigate whether there are other insects besides bees that feed on nectar and pollinate plants.
- Read more about honeybees' (or other insects') feeding behavior as it affects pollination.

## PART III

# Environmental Issues

Our interest in the well-being of our planet has surged and waned over the past few decades. The 90s appear to be the first decade in which the interest is more than a passing fad. People do not usually think of insects when they think of environmental problems, but insects play vital roles in almost every ecosystem and are affected by most forms of environmental stress. If two out of every three identified species is an insect and if there are more insects than all other forms of life on our planet, is it possible to think about our environment without thinking about insects?

Besides being a major component of most ecosystems, insects are also excellent organisms to use for studying environmental problems. Just as the canary was used to indicate the presence of dangerous gases in coal mines, insects can warn us of impending environmental stress. Two of the projects in this section will study how human-induced environmental stress—such as automobile pollution and oil slicks—affects insects. The other two projects use insects as experimental organisms in order to investigate the effects of human-induced environmental stress, including electromagnetic radiation and passive smoke.

# 8

# Electromagnetic radiation

## It's all around us

*(Does low-level EMR harm fruit flies?)*

Electrical circuits create not only electric fields but also magnetic fields. Both of these fields radiate from the devices containing the circuits, such as personal computers, hair dryers, television sets, electric blankets, and high tension wires. (See FIG. 8-1.) The magnetic fields created by electric circuits are in many ways similar to the magnetic fields created by magnets.

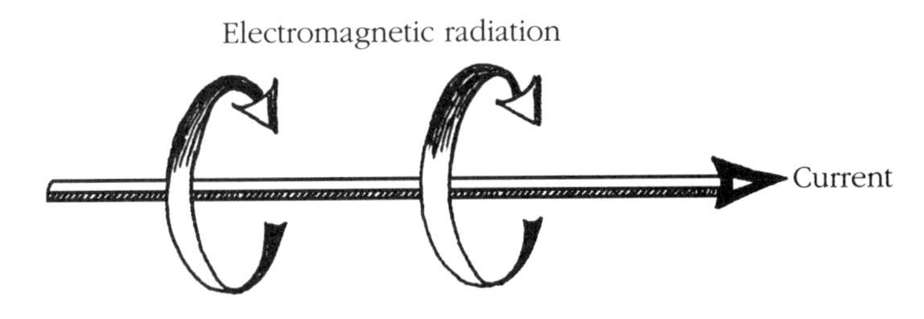


**Fig. 8-1** *Any electrical circuit creates a magnetic field surrounding the device.*

Both types of magnetic fields pass through many types of materials, become weaker the farther away they are from their source, and cannot be perceived by humans. The major difference between these two types of magnetic fields is that a magnet's field is constant, remaining in a fixed position to the magnet, whereas a magnetic field produced by an electrical circuit alternates direction. In the United States and Canada the circuit reverses itself 60 times per second (60 hertz). Fields that fluctuate at this speed are said to be in the extremely-low-frequency (or ELF) range.

Only recently have many of us become concerned about this radiation that surrounds us. Although highly controversial, some studies have shown a correlation between exposure to ELF radiation and such health risks as birth defects and childhood leukemia. This radiation is one of today's most debated health hazards.

## Project Overview

There is no debate as to whether people are routinely exposed to ELF radiation. Devices called gaussmeters are used to measure this radiation. Surveys can be done to see how much ELF radiation exists in various parts of your home or office. The question is whether these fields cause harm.

Epidemiological studies are being done in many countries to see if people routinely exposed to this radiation have a greater incidence of health problems. Studies that are done to determine if this radiation is harmful to other forms of life help us to know if these concerns are warranted. If lower forms of life are harmed by this radiation, we know that further studies are needed and that preventive measures should be taken, such as shielding oneself from the radiation (if that is possible).

The purpose of this project is to determine if fruit flies are affected by exposure to ELF radiation produced by television sets. Would exposure to the radiation produced by a television set increase morbidity (illness) or mortality (death) in fruit flies? Does exposure to this radiation affect the number of offspring produced? To answer these and any other questions your research leads you to, begin your literature search and formulate your hypotheses.

## Materials

- A Drosophila (fruit fly) culturing kit (These kits can be purchased from a supply house and include everything needed to rear generations of these flies. The kit may include all or some of the individual items listed below.)
- Four culture vials, plugs, and labels
- Instant Drosophila medium
- Measuring cup
- Sorting brush
- Anesthetizing fluid
- A pure culture of *Drosophila melanogaster* (You can use any strain but some are easier to sex than others. If you are ordering a culture, ask for the easiest strain to sex.)
- Dissecting scope for observing fruit flies

- White poster paper
- Two thermometers
- Two television sets

## Procedures

This project can be performed at any time of year. You must first procure a culture of a single strain of fruit flies. Place two tables at least 5 feet apart. They should be in the same amount of sunlight and in similar environmental conditions. One of the tables must have access to an electrical outlet. Place white posterboard on both tables so that color can be discounted as a variable. Place the two television sets on one table with the back of one television facing the front of the other. (See FIG. 8-2.)

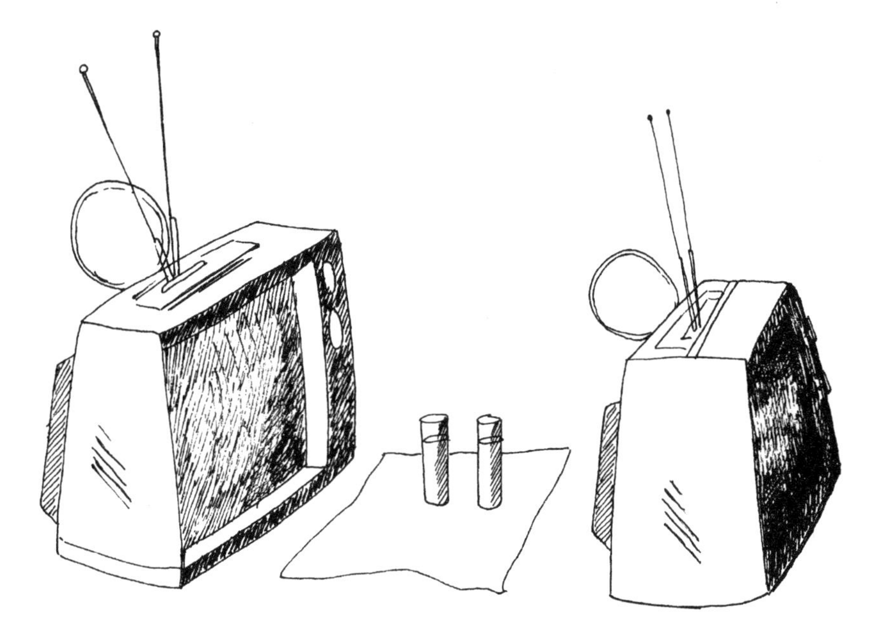

**Fig. 8-2** *Place the vials containing the experimental group of fruit flies 6 inches from each television set.*

Divide the culture of fruit flies into four culture vials containing an equal number of individuals. (Two will be used as experimental groups and two as controls.) Now, place two culture vials 6 inches from the front of one television and from the back of the other. Place the other culture vials on the control table containing no televisions. Place a thermometer on each table. Turn on the two televisions, turn off the volume, and make the picture black so that noise and the video are not variables. Begin the study early in the morning and note the date and time in your log. Leave the televisions on for 12 continuous hours and then turn them off for 12 hours. Repeat this process each day while examining both groups of vials for pupae. When you observe a substantial number of pupae in a vial (about 25), remove all the adults and record the date. (The control and experimental groups may not pupate at the same time.) (See FIG. 8-3.)

**Fig. 8-3** *Count the number of fruit fly pupae to be sure you'll have enough individuals for the next generation.*

Continue turning on and off the televisions each day for the vials containing only pupae. Observe each vial for emerging adults. When enough adults have emerged from each vial, remove and sex them using the dissecting microscope and collect six males and six females from each vial. Place them into a new culture vial. Place each new vial back on the proper table. Count any remaining individuals in the original vials and discard them. Take detailed notes of all observations. This will be generation #1.

Repeat this procedure for the next generation by waiting until enough pupae have formed in each vial, removing and sexing the adults, taking out 6 of each sex, and continuing the culture in a new vial. Don't forget to count the remaining individuals in the original vials before discarding. Do this until you've raised at least four generations of the control group. Enter your data in TABLE 8-1. (Each cycle takes about 16 days.) Average the results from the two experimental vials and average the results from the controls.

## Analysis

How many generations were raised in the experimental group when the control had completed four generations? How many offspring were raised in the control group as opposed to the experimental group? Were any forms of illness found in the control group? Did ELF radiation increase the morbidity, mortality, or overall life cycle of the flies? Compare the two groups by charting your results. If ELF radiation from a television has a negative effect on organisms such as the fruit fly, should you be concerned?

**Table 8-1**

| Generation # | INDIVIDUALS PRODUCED IN EACH GENERATION | | | |
|---|---|---|---|---|
| | Unexposed | | Exposed | |
| | Males | Females | Males | Females |
| 1 | 48 | 54 | 23 | 25 |
| 2 | | | | |
| 3 | | | | |
| 4 | | | | |
| | MORTALITY OF THOSE REARED | | | |
| | Lived | Died | Lived | Died |
| 1 | 11 | 1 | 8 | 3 |
| 2 | | | | |
| 3 | | | | |
| 4 | | | | |

## Going Further

- Perform the same experiment, but measure the actual ELF radiation that the flies were receiving. To do this measurement, procure a gaussmeter, which can be purchased or rented from Fairfield Engineering ([515] 472–5551), among other places. Take readings at the exact location where the experimental culture was placed on the table.
- Take readings throughout your home to see what areas are soaked in ELF radiation and try to figure out ways to minimize the radiation. Be sure to take readings around a personal computer.
- See *Environmental Science: High School Science Fair Projects* published by TAB/McGraw-Hill (#4515), for additional experiments on ELF radiation.

## Suggested Research

- Research the latest studies of the effects of ELF radiation. New studies are being published all the time.
- Research whether screens for a personal computer monitor reduce the amount of magnetic radiation. Be sure not to confuse electrical radiation with magnetic radiation. Are the marketing claims made about these screens deceiving?

# 9

# Passive smoke
## A universal problem

*(Does passive smoke harm mosquito larvae?)*

The negative effects of smoking on health have long been known. However, the dangers of passive smoke (also called environmental tobacco smoke) however have only recently been under attack. Passive smoke refers to smoke inhaled, not by the smoker, but by those around the smoker. Epidemiological surveys and studies have been undertaken to investigate whether people exposed to this smoke experience ill effects.

Of course, actual experiments cannot be performed on humans, so research focuses on laboratory animals. Even though the negative effect of a stimulus (smoke) on one type of organism doesn't necessarily mean that the same effect will be seen on another organism, it does put up a warning flag indicating that further research is necessary.

### Project Overview

Insects don't have lungs and don't use blood to carry oxygen to cells throughout the body. Instead, they have a tracheal system, or a series of interconnecting tubes running through the entire body. These tubes, called trachea, reach the outside of the insect's body through openings called spiracles. The system of tubes carries air deep within the insect's body. The tubes divide and get smaller and smaller until they finally reach groups of individual cells.

Aquatic insects, both adult and immature forms, have specialized structures that allow the spiracles to obtain oxygen while the insect is underwater. Some adult aquatic insects hold an air bubble up against their body as they dive, allowing the air to pass through the spiracles even while underwater. Mosquito larvae live just beneath the water's surface, and most have special breathing tubes that pass through the water's surface and allow air to enter. A pollutant like an oil slick can be disastrous to mosquito larvae since their oxygen lifeline is cut off at the source.

How would passive smoke affect mosquito larvae? Could the larvae continue to breath? If so, would smoke have any serious effect on the health of these immature forms? Would it affect their likelihood of becoming adults? To answer these and any other questions that your research leads you to, begin your literature search and formulate your hypotheses.

## Materials

- 3½ feet of plastic tubing (½-inch diameter)
- Two plastic or rubber food containers about quart size with airtight seal covers (You'll have to put holes in the covers.)
- Hot-glue gun for sealing the holes
- About 50 mosquito larvae (You can collect these in the summer by placing a bowl of water containing a few dead leaves outside for a few days, or the larvae can be ordered from a supply house. If you order the larvae, get either *Aedes* or *Anopheles*, not *Culex*.)
- Mosquito food (Either order this food when your order the mosquitoes or use flake goldfish food that you've crushed into a dust.)
- A few cigarettes with filters
- Matches
- Cotton
- Two small pieces of mesh, such as screening
- Packing tape or small rubber bands
- Magnifying glass
- Utility knife
- Marker
- Small piece of aluminum foil

## Procedures

This project can be performed at any time of year if you order the mosquitoes from a supply house. If you plan to collect your own mosquitoes, however, you must do it during the warm weather. The project is divided into two parts. The first part involves the preparation of the apparatus, and the second involves the experimentation.

Take the utility knife and into the top of each bowl cut two holes big enough to fit the tubing through. Cut the holes at opposite ends of the top and about 1 inch from the edge as shown in FIG. 9-1. Cut the plastic

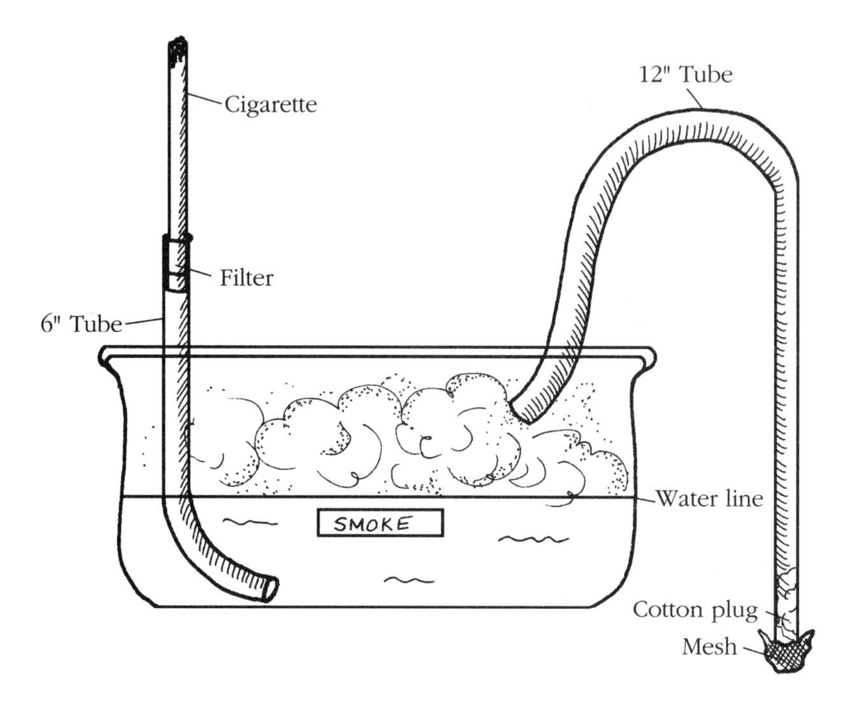


**Fig. 9-1** *Use quart-size rubber tubs to build this apparatus.*

tubing into two 12-inch lengths and two 6-inch lengths. Insert a 6-inch length into one of the holes and pull it through so that it almost reaches the bottom of the bowl. There should be only an inch or two sticking out of the bowl.

Next, insert a 12-inch length into the other hole. It should extend into the bowl only about 1 inch, with the remainder sticking out of the bowl—that is, there should be about 11 inches sticking out of the bowl, as shown in FIG. 9-1. When the tubes are properly positioned, use a hot-glue gun to form an airtight seal around the tubes and around the top of the bowl. Repeat this procedure for the second bowl. Mark one of the bowls "smoke" and the other "control."

If you ordered mosquitoes, they will probably come in the form of eggs. You want to begin the experiment as soon as the eggs hatch into young, very small larvae. If you collected the mosquitoes, be sure to begin the experiment as soon as the young larvae appear wriggling around in the water. (See FIG. 9-2.) Put about 1½ inches of pond water into each bowl. If pond water isn't available, use tap water that has sat for at least 24 hours. Be sure the 6-inch tube reaches well into the water and that the 12-inch tube remains well above the water, as shown in the illustration. Place about 25 young mosquito larvae into each bowl, add a pinch of food, and cover both bowls.

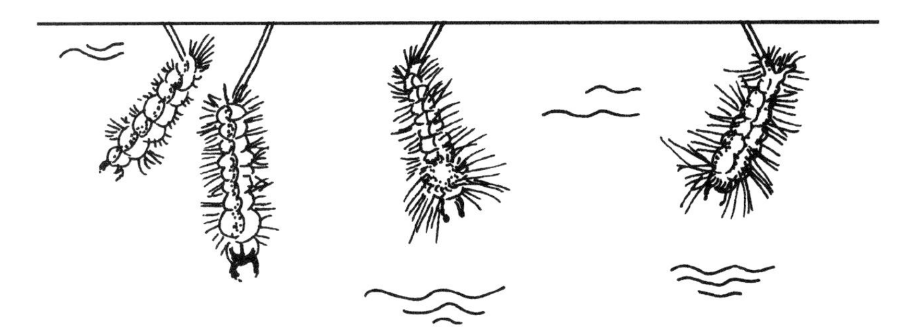

**Fig. 9-2** *You will see very small mosquito larvae wriggling about in the water.*

Place small wads of cotton into the ends of the 12-inch tubes. Pack the cotton tight enough so that it can't be easily forced out. Cover the tube opening (containing the cotton plug) with a small piece of mesh and secure it with a rubber band or tape wrapped around the tube, as shown in FIG. 9-3. (This is the tube that you'll be drawing air through, so make sure that the cotton isn't sucked into your mouth.) Most of the smoke will be caught by the cotton, so that it won't enter your mouth but will be trapped in the bowl. (The cotton is inserted into both bowls for consistency.) Insert the filter end of a cigarette into the end of the shorter tube on the bowl labeled "smoke." Wrap a small piece of aluminum foil around the bottom of the cigarette to catch ashes when they fall. Leave this tube open in the "control" bowl.

You are now ready to begin the experimental phase of the project. Light the cigarette and inhale through the long tube that contains the cotton plug. As you inhale, the lit end of the cigarette will brighten and smoke will be carried through the small tube, will bubble through the

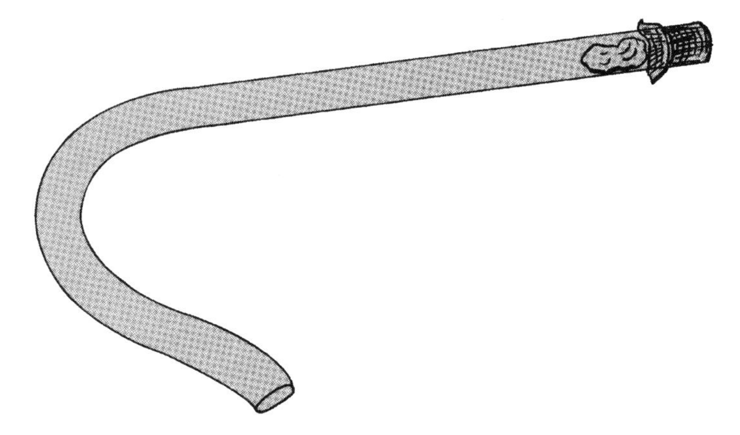

**Fig. 9-3** *At the end of the 12-inch tubes, insert a cotton plug and hold in place with a small piece of mesh.*

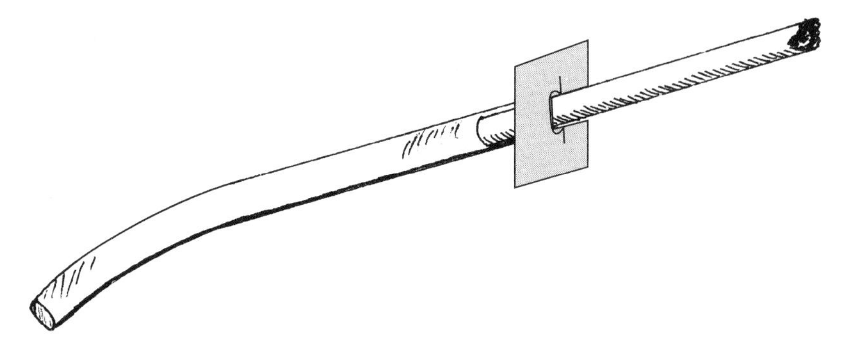

**Fig. 9-4** *At the end of the 6-inch tubes, insert the filter of a cigarette and place a small piece of foil to catch any ashes.*

water, and will enter the long tube through which you're inhaling. Most of the smoke, however, will be stopped by the cotton plug and trapped in the bowl.

CAUTION: Almost all the smoke will be stopped by the cotton plug, but do not inhale any smoke that might pass through the cotton.

Once you're sure that the apparatus is working properly, inhale through the tube 10 times. Then remove the lit cigarette from the tubing and dispose of it safely. The air in the bowl should now be filled with smoke.

Repeat the inhaling procedure with the "control" bowl, but without any cigarette. Only air will pass through the water. You'll still see bubbles, but there won't be any smoke. Inhale 10 times, as you did with the "smoke" bowl. Take notes. Leave the bowls undisturbed for 24 hours and then repeat this inhaling procedure. Continue these steps for 7 days. Feed the larvae once or twice each day. Use the magnifying lens to observe the larvae.

Look for any dead larvae in both bowls. Look for larvae that have pupated. Once you find pupae, continue the experiment to see how many emerge as adults. When the adults have emerged, release them outside.

## Analysis

How did the passive smoke affect the mosquitoes? Was there more mortality in the "smoke" than in the "control" bowl? Did the larvae show any ill effects? Did the same number of mosquito larvae pupate in both bowls? Research how the smoke entered the mosquitoes' bodies and how it might have affected them.

## Going Further

- Devise an experiment that would study whether passive smoke affects the adult mosquitoes after they have emerged?
- Devise an experiment to study the effects of passive smoke.

## Suggested Research

- Survey how aquatic insects breath underwater.
- Find the latest research on the dangers of passive smoking. What organisms are being used for these experiments?

# 10

# Oil slicks
## Endangered ecosystems

*(The effect of oil slicks on insects and ecosystems)*

Not only do insects affect our planet's environment, but environmental stress often affects insects. Some insects are even on the list of endangered species. A few butterflies are on the verge of extinction because the plants that they need for survival are being lost to real estate development in rural areas. Certain species of tiger beetles and dragonflies are on the list of endangered species because their habitats have been greatly reduced.

Although only a few insects are in danger of disappearing from the face of the earth, many insects are affected by pollutants that we produce daily. Aquatic insects are particularly sensitive to pollutants in streams and lakes. Humans have historically considered streams and rivers as natural toilet bowls, and many people still think this way. There are many sources of aquatic pollution: industry, farms, highways, and cars, to name a few.

### Project Overview

Many insects live on or in fresh water. Many spend their lives at the water's surface searching for food while others live deep within the water most of their lives. Water has a physical property called surface tension, which is the force created by the cohesiveness of water molecules. Very small organisms usually cannot escape this tension and become trapped in water. Some larger insects, however, depend for their survival on the surface tension of water. Water striders, whirligig beetles (see FIG. 10-1), and springtails are all insects that live on top of water and use surface tension to help them travel over the water's surface while searching for food, while mating, or otherwise surviving.

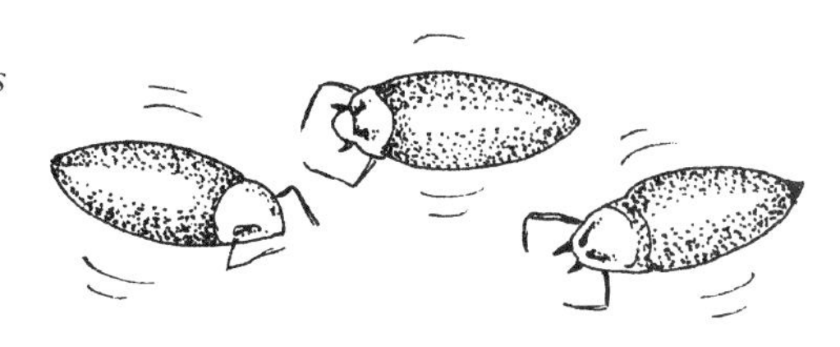

**Fig. 10-1** *The whirligig beetle is a typical aquatic insect that uses the water's surface tension.*

Contaminants in water, such as oil, affect the water's surface tension. A common form of pollution is the trail of oil left on the water's surface by motor boats. Does an oil slick affect aquatic insects? Is this form of pollution a simple annoyance or a life-threatening dilemma? What happens to aquatic insects that normally travel on top of water when the water's natural surface tension is destroyed by this form of pollution? Can these aquatic insects survive? Other insects, such as mosquito larvae, live just beneath the water's surface where they feed and breath. How would an oil slick affect their lives? Can they survive this type of pollution? To answer these and any other questions your research leads you to, begin your literature search and formulate your hypotheses.

## Materials

- At least 15 mosquito larvae (These can be collected from a pond or from the water left in a container or old tire for a few days during the summer. If the larvae can't be found, you can order them from a supply house.)
- Two or three live insects that inhabit the water's surface (Using a sieve or strainer, you can collect these insects during warm months or order them from a supply house. Water striders (*Gerris marginatus*), which are also called pond skaters, and whirligig beetles (*Dineutes americanus*) are easy to find. Others include the broad-shouldered water strider, water treaders and marsh treaders. You should have two or three of each type.

⚠ CAUTION: Many aquatic insects can inflict a painful bite. Don't handle them with your bare hands!

- Eyedropper or bulb pipette
- Measuring cup
- Flake fish food (from a pet store)
- Three or four similar containers (You may use cut-off, 1-quart milk cartons, which should all have the same amount of surface area when filled.)
- 3 tablespoons of vegetable oil (which will mimic motor oil)

- ½- & 1-teaspoon measurers
- Nylon material
- Rubber bands
- Measuring cup
- Dissecting microscope or magnifying glass

## Procedures

If you purchase all the insects, this project can be performed any time of year. During warmer months, however, it should be easy to collect all of the insects needed. In the first part of this experiment, you will examine the effect of an oil slick on mosquito larvae that live just beneath the water's surface. In the second part of the experiment you will observe the effect of an oil slick on adult insects that live on the water's surface.

For the first part, fill each of the three containers with 1 cup of water (use the same water from which the mosquitos were collected). Take a very small pinch of the flake fish food, grind it up in your fingers, and drop it into the first container. Do the same for the other two containers.

Label one container "control," which will have no oil. Label the second container #1. Cover about half of the water's surface with oil by placing about ½ teaspoon of oil and adding more oil until the surface is half covered. Label the next container #2. Use the eyedropper to place 5 mosquito larvae in each of the three containers. Then, in container #2, cover the water's entire surface with oil (see FIG. 10-2) (start with 1 teaspoon of oil and add more until the water's entire surface contains a thin

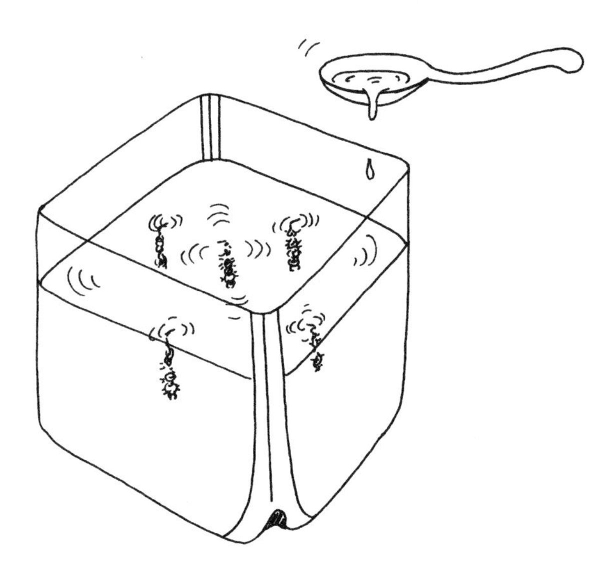

**Fig. 10-2** *Place oil on the water's surface to see its effect on the mosquito larvae.*

film). Record the date and time when this procedure was completed and make careful observations about the behavior of the larvae in all three containers. After your observations are complete, cover the containers with nylon material and secure them with rubber bands.

Each day look for emerging adult mosquitoes before removing the covers (they will be flying around in the container, under the netting). You can release them outside once they have been counted. Whether adults have been found or not, remove the covers each day to make observations. Look for mosquito pupae that are the intermediate stage between larvae and adult. Count the number of larvae, pupae, and adults that are present each time and also observe the condition of the larvae. Continue this procedure until adults are no longer found or until the larvae have died. Organize and analyze the data from all three containers to see how the oil affected the development of the mosquitos.

For the next part of the experiment, use the same containers (after you've cleaned them) to examine adults that swim on the surface. Put 2 cups of water in each container and add one kind of aquatic insect to each container (you can put two or three of the same type of insect into each container). Observe the natural movements of the insects for at least 10 minutes. Carefully note how each insect touches and moves across the water. Record your observations.

Then, add a few tablespoons of oil to each container until the water's surface is completely covered. A thin film covering the entire surface is all you need. Once again, observe how the insects in each container touch the water. How is their movement or behavior different than it was before? Record your observations.

## Analysis

In the first part of the experiment, what happened to the mosquitoes and why? Use a dissecting microscope or magnifying glass to look at the anatomy of a mosquito. (If you don't have a microscope, find a diagram of the anatomy of a mosquito larva. What anatomical structures are important in this experiment and why? Comparing the larvae in the three containers, what differences in larvae behavior did you see? How many adults emerged out of each container? Can the mosquitoes successfully pupate and emerge into adults if part of the water's surface is covered with oil? If all of the water is covered with oil? Did the oil affect the mosquitoes' development?

In the second part of the experiment, how did the insects move in the unpolluted water? How did they move in the oil slick? How did the movements differ? Could these insects survive in this polluted environment? Research what happened to the surface tension and what affect it had on the insects.

How does an oil slick, commonly found in navigable waterways, affect the aquatic insects' life cycles and behavior? What ramifications would your findings have for the ecology of the rest of the region, if any?

## Going Further

- Continue the second part of the project by keeping the insects for a few weeks in order to observe what happens to them under these environmental conditions. You'll have to read about what these insects eat so that you can supply food. How long can the control group survive compared to the polluted groups? What are the ramifications in actual ecosystems?
- Remove the surface insects from the polluted water and place them back in normal water. How long does it take for them to recover? Can they recover or have they incurred permanent damage?

## Research Suggestions

- Research how other forms of human intervention affect the lives of insects?
- Research "bioremediation." Are any insects used for bioremediation purposes?

# 11

# Road pollution

## Biodiversity & the automobile

*(The effects of road pollution on populations of insects and other arthropods)*

Sources of pollution are often categorized as either stationary or mobile. The most important stationary sources of pollution are coal and oil power plants that generate electricity and spew forth pollutants. The primary mobile source of pollution is the automobile with its gasoline driven engine. Automobiles are the major contributor of carbon monoxide, hydrocarbons, and nitrogen compounds.

In the past 25 years there have been major technical advances that have helped to reduce tailpipe emissions, but unfortunately the problem continues to worsen rather than improve. Even though the emissions released from each car are less than before, there are far more cars—with the net result that there is more, not less, pollution. The pollution caused by cars not only contributes to global warming and ground-level ozone (smog) but also to a general decline in local ecosystems.

## Project Overview

The habitat near well-traveled roads is dramatically altered by tailpipe emissions and other automobile by-products, such as oil. In addition, the spreading of large volumes of sand and deicing salts, in some parts of the country, affects the area adjacent to these roads. Salts and sand can kill many indigenous plants. High concentrations of carbon dioxide and carbon monoxide can have a dramatic effect on which insects can and cannot live in these areas. In turn, changes in these insect populations affect higher trophic levels—and, thus, the entire local ecosystem.

This project will study how automobile traffic affects roadside ecosystems. Is there a difference in the populations of insects and other arthropods (see glossary) that live in proximity to the road and those that live farther away? Is there a gradual increase in biodiversity as you go farther away from a road? Do some arthropods prefer to live near a road? To answer these and any other questions that your research leads you to, begin your literature search and formulate your hypotheses.

## Materials

- Insect-collecting net (preferably a sweep net). (See chapter 1 for a description of sweeping and sweep nets.)
- 12 glass jars, with tight caps for holding liquids
- Markers with which to label the jars
- Meter stick or measuring tape
- Six stakes about 3 feet long
- Hammer or mallet to drive the stakes into the ground
- Scissors and ball of thin string
- A well-traveled road with several yards of flat open field on both sides (It is not necessary that there be heavy traffic when the experiment is performed. Pick a time when the volume of traffic is low. See warning below.)
- A few white pans to dump insects into for identification
- Fine forceps
- Magnifying lens or dissecting scope
- Insect identification guide

## Procedures

CAUTION: This project must be performed under the strict supervision of your sponsor. Since it involves activity along a roadside, the project can be dangerous if extreme caution is not taken. Investigate where this project can be performed safely. A road with a wide shoulder will provide an extra margin of safety. Pick a time during the day when the road is least traveled. If possible, select a road that has recently been closed to traffic for a day or two.

Prepare 12 jars by labelling each with one of the following: "1MA," "2MA," "3MA," "1MB," "2MB," "3MB," "1MC," etc. (the "M" stands for the number of meters from the roadside and the letter represents the sector). The result will be four sets containing three jars each. To preserve the insects, fill each jar with ½ inch of 70% rubbing alcohol. Bring these jars to the roadside that you and your sponsor have selected.

CAUTION: Before going to the roadside, be sure to wear protective clothing. Be aware of ticks, especially in areas where Lyme disease is a problem. Consider wearing an insect repellant containing DEET as the active ingredient. Be sure, too, that the roadside area you've chosen does not contain thorny vegetation that will make sweep netting difficult.

Drive into the ground two stakes 10 meters apart and 1 meter away from the edge of the road's shoulder. Tie a length of string between these stakes, as shown in FIG. 11-1, so that the string is parallel with the road. Sweep net along this strip of land. (See the Introduction, chapter 1, for instructions on how to use the sweep net.) Make 10 passes back and forth over this strip of land that runs parallel with the road. When 10 passes are completed, place the entire contents of the net into the jar labelled "1MA."

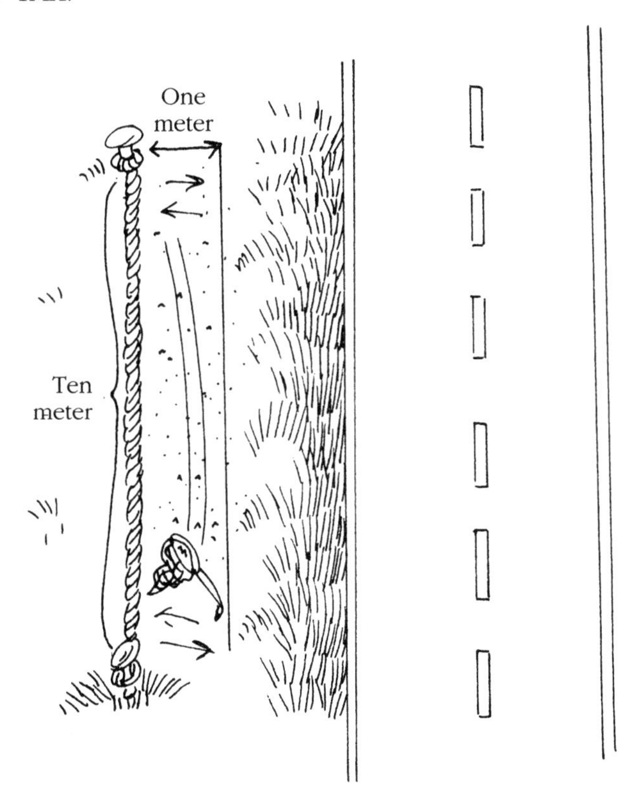


**Fig. 11-1** *Place the stakes and the string horizontally, one meter from the roadside shoulder to create the first sector.*

Next, measure another 1-meter strip running parallel to the first string. Pound in two more stakes and tie a string between them (these stakes are now 2 meters from the roadside). Once again, use the sweep net along this strip and place the contents in the jar labelled "2MA." Create a third strip beyond the second line of string, repeat the sweep-net procedure, and place the contents in the jar labelled "3MA." You now have a complete set of sweeps at 1, 2 and 3 meters from the road's shoulder in sector "A."

To increase the validity of your results, create three more sectors of three strips each, as shown in FIG. 11-2. You can either create two sectors on each side of the road, totalling four, or create all four on the same side of the road adjacent to each other.

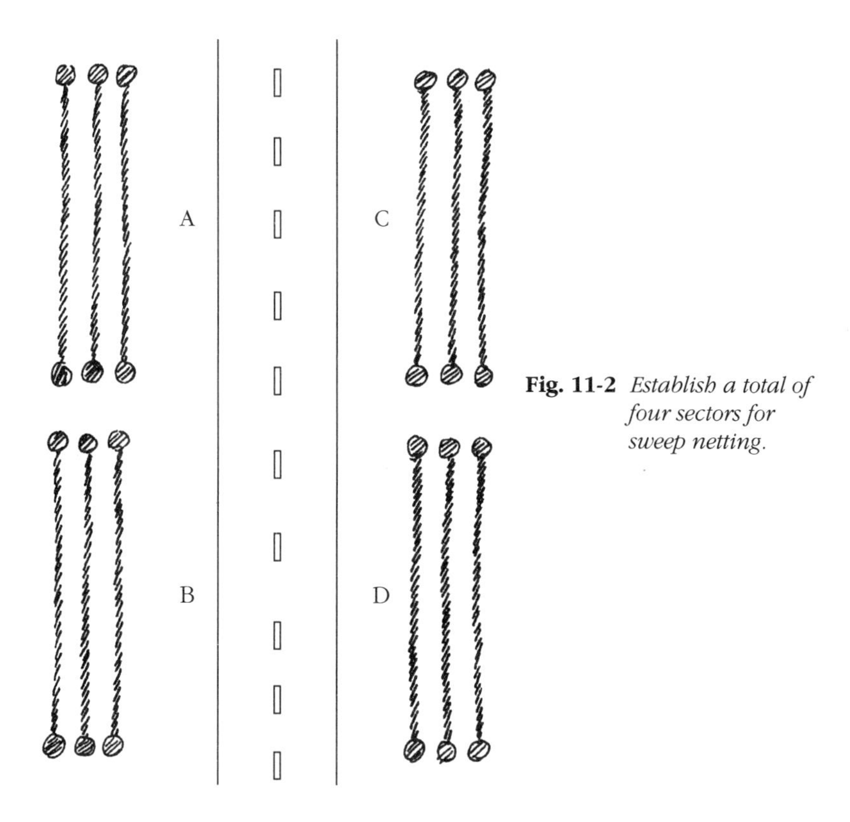


**Fig. 11-2** *Establish a total of four sectors for sweep netting.*

When collections from all four sectors have been in the jars, remove all the stakes and strings and return to your home or lab to study your collections. (*Be sure to check yourself for ticks before continuing the project.*) Dump the contents of each jar into a white pan to identify the catch. Identify as many of the insects and other arthropods as possible (you'll need a magnifying lens or dissecting microscope to do this). You don't have to identify each specimen down to the species. For insects just identify to the order, such as flies (*Diptera*), beetles (*Coleoptera*), bugs

(*Hemiptera*); and for other arthropods just identify the major groups, such as spiders and ticks. Keep the jars separate and keep a log of what were found and where they were found. Be sure to also count the numbers of each group found. Record your data in TABLE 11-1.

**Table 11-1**

| Sectors | Ticks | Spiders | Beetles | Flies | Grasshoppers | Others | Totals |
|---|---|---|---|---|---|---|---|
| 1MA | 1 | 0 | 0 | 0 | 11 | 1 | 4 |
| 2MA | | | | | | | |
| 3MA | | | | | | | |
| 1MB | | | | | | | |
| 2MB | | | | | | | |
| 3MB | | | | | | | |
| 1MC | | | | | | | |
| 2MC | | | | | | | |
| 3MC | | | | | | | |
| 1MD | | | | | | | |
| 2MD | | | | | | | |
| 3MD | | | | | | | |

When the identification and counting is done, average all four counts from the 1M, 2M, and 3M strips to get your final numbers.

## Analysis

What types of arthropods did you collect at each meter strip from the roadside? Did the strips differ in the diversity of organisms found in them or in the numbers of each type found? Did a gradient exist as you went farther from the road? If so, what factors made it so? Were there any types of organisms that preferred the polluted area near the road's edge; if so, can you determine why? Graph your results illustrating how the populations changed as you went farther away from the road.

If differences were found, how would the ramifications be for the entire local ecosystem? How might these effects be considered warnings of larger-scale environmental problems?

## Going Further

- Continue this project by examining aquatic ecosystems that are adjacent to roads.
- Perform similar experiments near other sources of pollution, such as areas around power plants or beneath high-tension wires.

## Suggested Research

- Investigate why some organisms were and some were not found along the road. Were oil by-products, carbon monoxide and dioxide gases, or road salts responsible for your results.
- Look into other negative environmental effects caused by highway traffic.

## PART IV

# Insect Ecology

Ecology is the study of the relationship between organisms and their environment. By environment, we mean both nonliving things—such as weather conditions and soil type—and living things. Because of the incredible number of insects and insect species, insects affect, directly or indirectly, all life on our planet. Therefore any factor that affects insects will most likely affect the entire ecosystem that they inhabit. A thorough understanding of how insects fit into intricate food webs is crucial to understanding how ecosystems exist.

Insects form the lower trophic levels of many food chains acting as a food source for secondary consumers. Many plants are totally dependent upon insects for pollination. Insects are also important in recycling nutrients, thus providing a continuous source of chemical energy for almost all ecosystems.

Insects affect the environment in many ways, but the environment can also affect the insects. Insects are affected by other organisms and especially by human activities. Although pest control is essential to assure an adequate food supply, unwise use of some pest-control methods, such as the exclusive and rampant use of synthetic pesticides, can cause more harm than good. This is especially true when less harmful—but effective—methods, such as integrated pest management, are available. When we pollute the environment and speak of the effect of such pollution on the ecosystem, it is often insect populations that are first affected, resulting in a domino effect and a collapsing ecosystem.

The following general ecology textbooks have in-depth coverage of insect ecology: Price (1984) and Huffacer and Rabb (1984). For lighter reading on insect ecology, try *Insect Life*, by Frost, or *Insects, Food, and Ecology* by Charles T. Brues. Although both books are dated, they are informative and enjoyable to read.

# 12

# Biodiversity

## Habitat destruction

*(Habitat destruction and the biodiversity of insects)*

There is a great deal of debate about a continuous decline in our planet's *biodiversity*, the different kinds of organisms that exist on our planet. Today, roughly a million and a half species of life forms of have been identified. Estimates of how many species actually exist range from 10 to 100 times that number.

Human activity has been forcing many organisms into extinction and has been endangering many others. Human activity takes many forms, but the most common is the destruction of an organism's habitat. Urbanization and large-scale agriculture can push a species out of a region, if not off the planet completely. It is estimated that for every 10% of a natural habitat lost to development, 50% of its species are also lost.

## Project Overview

Of the million and a half identified species of life on our planet, almost 1 million of them are insects, and about two thirds of those insects are beetles (*Coleoptera*). (See FIG. 12-1.) As mentioned previously, human activity affects all forms of life, some much more visibly than others. When people speak about the loss of our planet's biodiversity, they are usually talking about higher forms of life, but the importance of insects cannot be denied. Humans take wild habitats containing many species of plants and produce monocultures that contain a single species, such as turf or a crop, thereby altering the accompanying biodiversity.

Clearcutting forests in the Pacific Northwest has put the spotlight on the spotted owl, but what does such cutting do to the populations of insects in the region? How do other forms of habitat destruction affect insects? Since most insects fill the lower tiers of a food chain, how does their position in the food chain affect the entire ecosystem of a region?

In this project you will survey insect populations in various habitats, both natural and altered. Your goal will be to determine the relationship between habitat destruction and insect populations.

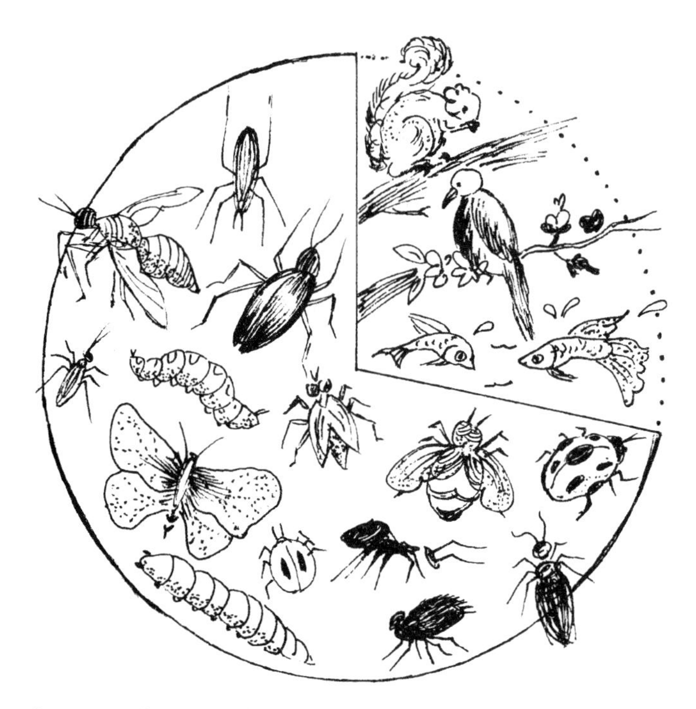

**Fig. 12-1** *About two-thirds of all the species identified are insects.*

Do insect populations dramatically change in diversity and in numbers when we alter their habitat? How great is the effect of this change in their habitat? Is the change always the same regardless of the type of ecosystem? To answer these and any other questions your research leads you to, begin your literature search and formulate your hypotheses.

## Materials

- Access to at least four areas, including: (1) a tended, uniform grassy area (such as a lawn or golf course), (2) an untended field or meadow with at least knee high growth, (3) a forested area, and (4) a cleared part of a forest (such as a portion adjacent to the trees that have recently been cleared for development or agriculture)
- Sweep net (See introduction in chapter 1.)

- Four killing jars and sufficient activating fluid (See introduction in chapter 1.)
- Collecting aspirator (See introduction in chapter 1.)
- Marker for labeling the jars
- Heavy white paper or a white pan for counting insects
- Fine forceps for manipulating insects during identification

- Insect field guide
- Dissecting scope or quality magnifying lens

## Procedures

Charge the killing jars with activating fluid before you go collecting. Label each of the jars so that you can identify the collection: for example, "lawn," "field," "woods," and "clearing" (See FIG. 12-2.) Since different areas will require different collecting techniques, you should collect within a specified area at each site. Go to either the lawn, field, or clearing to begin collecting. Make 10 sweeps over the first site (see chapter 1 for sweep-net instructions). Dump the contents you collect into the properly labeled jar. Repeat this procedure three more times, thus making a total of 40 sweeps at that site. Keep the area or radius you have swept consistent among all groups. Each time you sweep an area, dump the insects into the same jar.

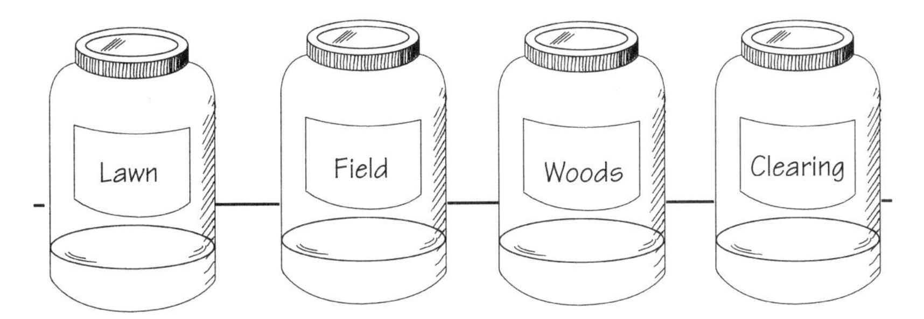


**Fig. 12-2** *Label each jar with the area where you performed the sweep.*

Move on to the wooded site. Since sweeping will not work well in this type of habitat, use a net and collecting aspirator to collect individual insects from trees and other vegetation. To remain consistent among sites, collect insects within the same area or radius as that at the first three sites. (Do not collect any insects from the ground since this was not done at the other sites.) Put the insects into the properly labelled jar.

Once you've collected from all sites, return to your home or lab. Check yourself for ticks before continuing the project. Before proceeding with the experiment, all the insects should remain in a jar for 2 hours. Dump the contents of the first killing jar onto a white piece of paper or white pan and separate the contents into general categories. Then use your insect identification guide to identify as many as possible. In addition, count the total numbers of each type. Repeat this procedure for all four jars. Record your observations in TABLE 12-1. When this is completed, compare the results of the tended lawn to the untended field or meadow, and the results of the woods to the cleared area.

**Table 12-1**

| GRASS (MONOCULTURE) RESULTS | | | |
|---|---|---|---|
| Sweep# | Order | Type of insect | New (Y or N) |
| 1 | Diptera | Midge | Y |
| | " | Housefly | Y |
| | Hymenoptera | Wasp | Y |
| 2 | Diptera | Midge | N |
| | Hymenoptera | Wasp | N |
| | Homoptera | Plant hopper | Y |
| | Coleoptera | Flea Beetle | Y |
| 3 | | | |
| 4 | | | |
| 5 | | | |
| 6 | | | |
| Total types of insects found in all sweeps | | | Diptera 2<br>Hymenoptera 1<br>Homoptera 1<br>Coleoptera 1 |

## Analysis

How do the results between the natural sites and the tended sites compare in diversity? Is there more diversity on the lawn or the meadow? Is there more diversity in the woods or in the cleared area adjacent to the woods? What factors do you think are affecting the biodiversity of these habitats? Do the same factors control both relationships?

## Going Further

- Determine which factor is responsible for changing the insect diversity in the project above. Devise an experiment to see if the amount of sunlight or the diversity of plant species in the area is the most important factor.
- Perform a similar experiment but collect from a large farm growing a single crop. (See FIG. 12-3.)

**Fig. 12-3** *A monoculture encourages large numbers of insect pests and requires large quantities of pesticides.*

## Suggested Research

- Research what happens to insect diversity when a tropical rainforest is clearcut. How does this habitat differ from those that were studied in the project above?
- Research the problems caused by large-scale monocultures and the ways in which we try to overcome these problems.
- Study extinction from a historical perspective. Is it rare for organisms to become extinct? Are humans always the cause of the demise of a species?

# 13

# Humus
## What goes around, comes around

*(The relationship between humus and the diversity and abundance of insects)*

Soil is a mixture of minerals (inorganic matter) and dead, decomposing organisms (organic matter). When organic matter is decomposed and mixes with inorganic matter, it is called humus. Different soils have different proportions of inorganic versus organic materials. Fertile topsoil, the upper level of soil, is often rich in humus.

Topsoil acts as a transitional region between the living and nonliving worlds. Minerals in topsoil are absorbed by plants and become part of the living world where they are passed along food chains to animals. As these plants and animals void waste products and die, they return the nutrients to the soil where they can be used again.

Topsoil is often teeming with life, most of which is hidden from the naked eye. One gram of soil can contain billions of bacteria, half a million fungi, and tens of thousands of algae and protozoans. In addition, there are often hoards of insects and earthworms inhabiting topsoil.

## Project Overview

Insects that live in the soil are found in great diversity and in vast numbers. It is possible to find thousands of insects in a few cubic inches of soil. Insects play a vital role in detritus (decomposing) food chains. They help decompose almost all forms of organic matter and thereby enrich the soil by helping to create humus. (See FIG. 13-1.)

Is there a direct relation between the amount of humus present in soil and the diversity and number of insects that live in that soil? Can the

**Fig. 13-1** *Insects such as carrion beetles help decompose organic matter and return it to the earth.*

diversity of insects be predicted by the amount of humus in an ecosystem? Does this relation differ depending on soil type? To answer these and any other questions that your research leads you to, begin your literature search and formulate your hypotheses.

The "Procedures" section of this project describes how to build a soil-insect collection device called a Berlese funnel. It would be advantageous to build these devices (or purchase them) before beginning this project.

## Materials

- Ten lunch-size paper bags
- Garden trowel
- Humus test kit (available at a scientific supply house or possibly an organic gardening center)
- Five quart-size funnels
- Five lamps with incandescent bulbs (30 watts)
- Five jelly jars (any 12-ounce glass jars)
- 5 cardboard boxes (2 inches to 4 inches higher than the height of the jelly jars)
- Window screening or something similar
- Knife with pointed tip
- Rubbing alcohol (70%)
- Magnifying glass or dissecting microscope

## Procedures

This project is best performed from early summer through early fall, but can be performed any time the ground is not frozen or covered with snow. You will be collecting five different kinds of soils and use a Berlese funnel to collect the insects inhabiting each kind of soil. (See FIG. 13-2.) In addition, you will be collecting soil samples at each location to test for the abundance of humus. Finally, you will be correlating your biodiversity findings with the humus-test results.

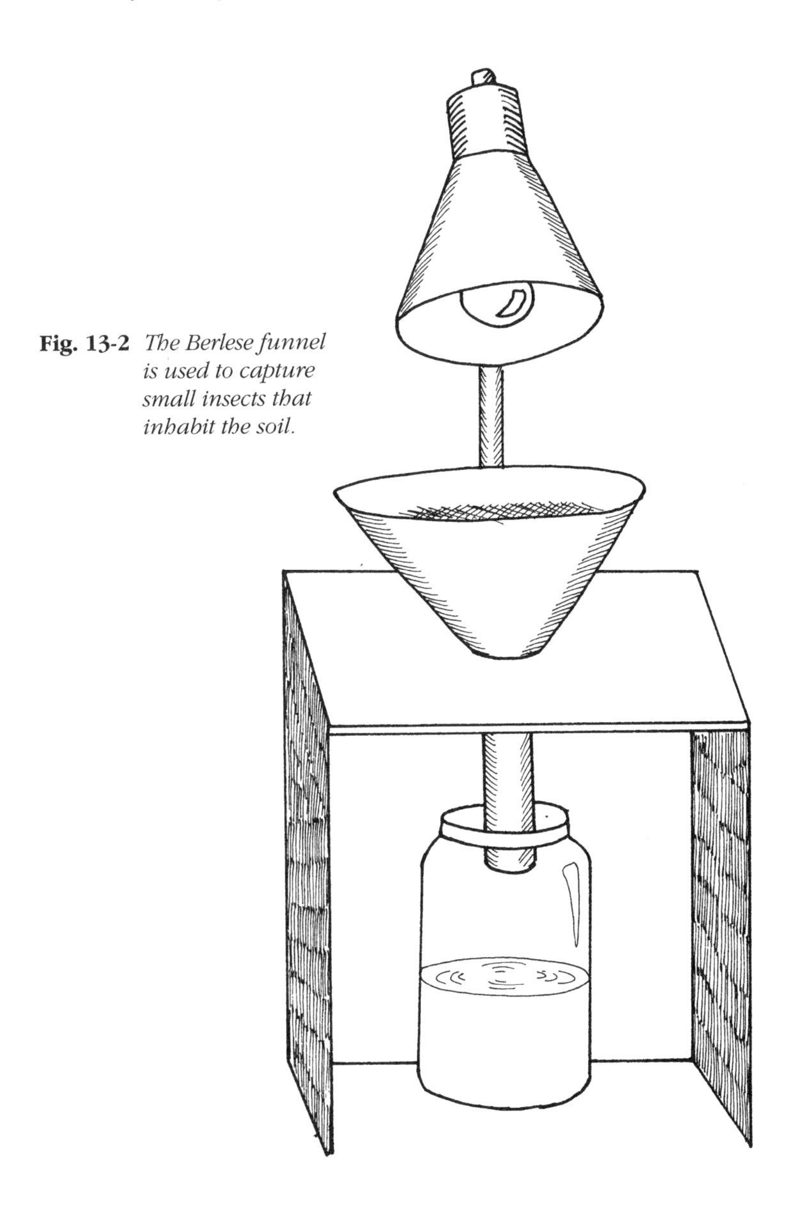

**Fig. 13-2** *The Berlese funnel is used to capture small insects that inhabit the soil.*

Locate areas containing the following kinds of soil: (1) a sandy soil; (2) a dark, fertile soil, such as that from a garden or compost heap, (3) a soil from the woods with a lot of dead, decomposing leaf litter; and (4) any other areas you are interested in testing and have access to. When collecting, take all the material from the top down—in other words, don't brush away the surface debris. In the woods, for example, collect the leaf litter as well as the surface soil.

At each site use the trowel to fill a lunch bag. (Consider collecting from three locations within each site in order to replicate your data collection.) Don't pack the material, just fill the bag. Close the tops of each bag tightly as soon as the soil is collected. Label each bag with the soil type and location as well as the date and time of collection. Mark these bags "funnel" to distinguish them from the next set of bags used for the humus test.

After collecting a sample (to be used for the funnel), collect soil for the humus test. Fill about one third of another bag at each site with soil. For these bags, you want to only include the actual surface soil. If leaf litter is present, brush it away and collect directly from the soil surface. Scoop up the top inch of soil and place it in the bag until it is one-third full (you need not be exact). Label each of these bags "humus" so that they can be matched with the proper "funnel" bags. (See FIG. 13-3.)

**Fig. 13-3** *Take two collections from each site—one for the funnel and one for the humus test.*

When you return home or to your school lab, test each of the samples marked "humus" with your humus test kit. (There are different types of tests, so follow the instructions that came with the kit or check with your sponsor if he or she provided the test.) Document the amount of humus present in each sample.

Next, set up the Berlese funnels, like the one shown in FIG. 13-2.

These are designed to collect insects living in soil. When the soil is heated up with the lights, the insects move deeper into the funnel to get away from the heat and eventually fall out through the bottom of the funnel and into a jar containing a preservative (alcohol). Berlese funnels can be constructed or be purchased.

To construct a Berlese funnel, cut a hole in the top of a cardboard box so that the funnel can rest in the hole. The height of the box should be designed so that the bottom of the funnel (its tip) just enters the jelly jar opening that sits beneath it. The end of the funnel should go about ½ inch to 1 inch into the jelly-jar opening (if it goes too far inside, it will be in the alcohol). (See FIG. 13-4.)

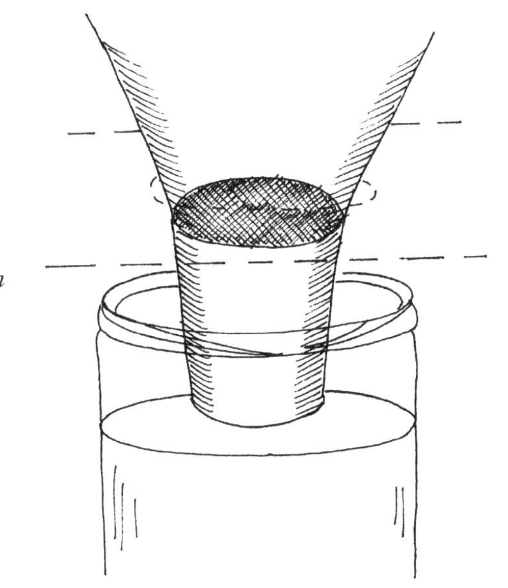

**Fig. 13-4** *The end of the funnel should be in the jar but above the alcohol.*

Put a layer of window screen (or similar mesh) inside the wide opening of the funnel, as shown in FIG. 13-4. Push the screen down slightly into the tip section of the funnel so that the screen remains in place. The screen keeps the soil from falling through the funnel but lets the insects pass through. Fill the jelly jar halfway with 70% rubbing alcohol. The alcohol kills and preserves the insects as they are collected. Place the jar under the tip of the funnel. Set up a lamp so that it's shining down on the top of the funnel. Label each funnel with the type of soil about to be placed into it.

CAUTION: Keep the bulbs a safe distance from the cardboard so that they are not fire hazards. The bulbs must be at least 2 inches away from the top of the soil. They must remain on for 1 to 2 weeks. At the beginning of the experiment, check the setup often to be sure there is no overheating problem.

Once the funnel is set up, you can add the soil. Move the lamp out of the way and slowly dump the soil out of the bag and into the funnel. Fill all the funnels with the same amount of soil. Some soil will fall through and into the alcohol, but most of it will stay in the upper part of the funnel. (If the soil is too sandy, consider placing an extra layer of window screening at the bottom of the funnel.) Put the lamps back in position. Do this for each funnel.

After the allotted time, turn off the lamps and remove the jars. (Most of the insects will fall out and into the alcohol within the first few days, so that you can actually begin identifying the insects while the funnels are still operating.) Use a dissecting microscope or a magnifying glass to separate the different types of insects and other invertebrates collected and to count the numbers of each type of insect in each collection.

Although it is not crucial to identify each type (since we are interested in how many different kinds were found), it would be helpful to create a collection of identified species for your project. Use your field guide and try to identify as many as possible.

## Analysis

From the first part of the project, how much humus was found in each type of soil? From the second part, how many different kinds of insects did you collect in each type of soil? What was the total number of insects collected from each type of soil? Are some soils better for a variety of insects to inhabit than others? Are some soils more capable than others of supporting a greater total number of insects?

Now compare the amount of humus in each soil type with the amount of insect biodiversity. Draw a graph that illustrates this relationship for each site. Is there an obvious relationship for any of the sites? Do you think insects are present in soil because of an abundance of humus or is the humus present in the soil because of the abundance of insects?

## Going Further

- Consider illustrating your data by creating an insect collection that not only shows the diversity of the insects you found in the tests but that also shows the relative numbers of each.
- You can do the same experiment but vary the depth of your soil collection. How do the insect populations differ at the surface, at 2 inches, and at 4 inches below the surface?

## Suggested Research

- Read more about soil insects. How are they beneficial to man and how are they harmful?
- Research whether soil insects have any effect on other important soil organisms, such as the earthworm, or on any of the nitrogen-fixing organisms found in the soil.

# 14

# Recycling
## Naturally

*(The role insects play in decomposition)*

Nutrients are absorbed by green plants (producers) and incorporated into their tissues. When animals (consumers) eat the plants and each other, these nutrients are passed along food chains and supply sustenance to all forms of life. When plants and animals die, their tissues decompose, and these nutrients are returned to the earth where they are recycled naturally.

This recycling of nutrients from the nonliving to the living world and back again is essential for the continuation of life on earth. If this recycling didn't happen, the nutrients of the dead organism could never be used again—and sooner or later, life on earth would run out of the basic building blocks of life. The earth would also be covered with mountains of dead plants and animals.

Food chains that describe feeding relationships in which complex organic molecules are broken down are called detritus (decomposing) food chains (as opposed to the more familiar grazing food chains in which complex organic molecules are built up). Most organisms that play a role in detritus food chains can be divided into two major categories: scavengers (which are primarily insects, along with some higher organisms) and decomposers (which are primarily bacteria and fungi).

## Project Overview

In almost all ecosystems, insect scavengers play important roles in detritus (decomposing) food chains (the major exception being aquatic ecosystems). Some organisms help break down only plant matter, others only animal matter. Others eat almost any decaying organic material. Some examples include carrion beetles that break down animal carcasses, dung beetle larvae that break down piles of dung, termites (and their intestinal guests) that break down woody material, and numerous insects that help turn leaf litter into duff.

In this project you will determine what roles insects play in decomposing organic matter in different types of ecosystems. The first part of this experiment will examine the importance of insects in the decomposition of meat. You will see how long it takes a piece of meat to decompose—with and without insects—in various habitats. In the second part you will compare the decomposition of wood in a variety of habitats. As agents of decomposition, are insects as important in an open field as they are in a forest, an ocean beach, or a freshwater pond? (See FIG. 14-1.) To answer these and any other questions your research leads you to, begin your literature search and formulate your hypotheses.

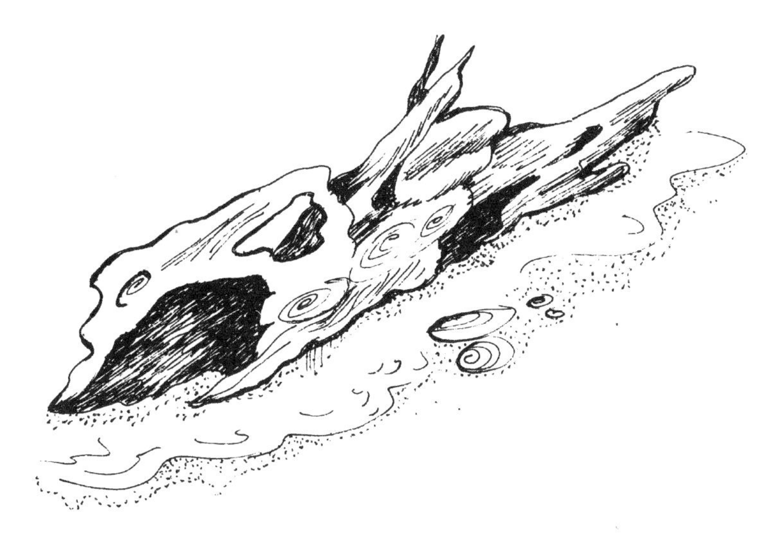

**Fig. 14-1** *Some organic matter decomposes more slowly than others.*

## Materials

- Access to three types of habitats, such as an open field; a wooded area; a dry, desertlike area; or a body of water
- Six similar pieces of meat (of similar size and containing similar amounts of fat and bone)
- Six jars (quart-size mason jars or 32-ounce mayonnaise jars)
- Nylon material
- Rubber bands

- Chicken wire or other strong mesh (small enough to block out mice and other rodents but wide enough to let in insects)
- Wire
- Wire clippers
- Stakes
- Hammer
- Three or four similar pieces of wood, such as cuttings from a dead tree branch
- Weatherproof spray paint
- Rubber gloves
- Safety goggles

## Procedures

This project should be done during the beginning of the warm-weather season (it can be performed any time of year, but will take much longer to complete in the winter). The first part of the experiment studies the decomposition of meat with and without insects in different environments. The second part studies the decomposition of wood in different environments, including those where the wood would not normally exist. To save time, both parts can be done concurrently. Or you might want to only perform the first part, which can stand alone as a science fair project.

For part one, put one piece of meat into each of the six jars. Mark three of the jars "insects," and the other three "no insects." Put a double layer of nylon material over the top of the "no insects" jars and secure the nylon with rubber bands. These jar covers will prevent insects from getting to the meat. The other jars will have no cover; therefore, insects will get in.

You must protect these jars from larger animals by building an "envelope" cage of wire mesh. (If you have a small, wire animal cage that is big enough to hold two mason jars, you can use it instead of building a cage.) To build a chicken-wire cage to protect the two jars, follow the instructions below.

Place the jars between two layers of the wire mesh. Tie the edges of the two layers of mesh tightly together with wire so that the jars cannot be pulled out by animals. Then place these cages containing the two jars in each of the three habitats. (See FIG. 14-2.) Try to find areas off the beaten path so that no one will stumble upon the cages. Be sure to keep the cages away from other person's homes and property because the cages will begin to smell and attract insects and other animals. Try also to keep the cages out of direct sunlight and on ground that is not too moist.

When the sites are found, stake the envelopes into the ground. Secure the cages firmly by hammering the stakes through the edges of the envelope and into the ground. Record the exact location of each site so

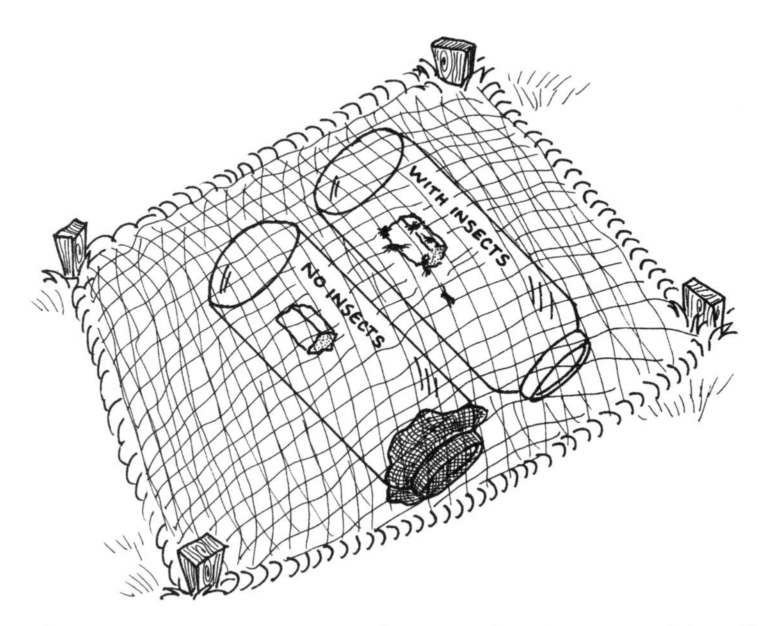


**Fig. 14-2** *The "envelope" cage protects the two jars from large animals but allows insects to enter.*

that you can locate the cages to make your observations. Record the date and time that each experiment begins. Every few days, return to these sites to record your observations for each jar in the cages. You can alternate your data collection between the sites, so that you are visiting each site once or twice a week but collecting data every day.

Make observations about the condition of the meat and the amount that is left. You do not need to touch the meat, only to observe it. What does the meat look like? Is it still solid? Are different portions (meat or bone) affected differently? Describe the presence of insects? When do you first see insects in the jars labeled "with insects"? (Hopefully, you won't see any in the other jar. If you do, the experiment must be rerun.)

When there is no further decomposition in the "with insects" jars, you can discontinue your observations. The length of time that you must continue collecting data depends on the habitat and the season. In warm months, this experiment will last at least a few weeks. In cold weather, it could last for many months.

At the end of the experiment, throw away the envelopes from each location.

 CAUTION: Wear rubber gloves and safety goggles while picking up the envelope and enclose it in a heavy-duty plastic bag. Close the bag tightly and dispose of it properly. Do not touch the decaying meat.

Organize and analyze the data collected for each site.

For the second part of this experiment, find three or four similar pieces of dead, seasoned wood. To be consistent, it is best to cut a branch of a dead tree into three or four equal pieces, each about a foot long. Place each piece in a different habitat. These habitats can be the same areas used in part one—but one of the areas should be heavily wooded, and one must be aquatic.

Before placing the pieces of wood at each location, mark them in some way so that you can identify them each time you visit a location: Spray paint a small area of each branch or place a nearby stick with a colored ribbon on it. These branches will probably have to remain in place for a couple of months. Place a piece of wood at each site. One piece, however, must be placed either on a sandy beach surrounding the water or in the shallow water of a pond, lake, or ocean.

Every few weeks, observe the wood at each site. Turn the pieces over to see if any insects are living on or in them. After making your observations, return each piece of wood to its original position. Continue these observations until at least one of the pieces has a substantial number of insects on or in it. Compare the other pieces in the other habitats. Do the pieces of wood in each ecosystem have insects living and feeding on them?

## Analysis

Study your observations. Compare the results for both sets of jars in the first part of the experiment. Did the meat decay differently—and, if so, how? Did the insects affect each portion of the meat (muscle, bone, fat) in the same way? Were insects important decomposers of meat in all the habitats? Record your data in a table similar to TABLE 14-1. What happened in the second part of the experiment? Did insects begin to decompose the wood in all the ecosystems? If not, why not? Did the wood placed on a sandy beach or in shallow water decompose as rapidly as in the other habitats?

## Going Further

- Devise an experiment to determine how a loss of biodiversity in an area affects decomposition. For example, how does organic matter decompose in a forest as opposed to an adjacent area that has been clearcut?
- To continue this experiment, find out if different types of meat (for example, beef vs. chicken) attract different types of insects.

**Table 14-1 SITE #1**

| WEEK # | INSECTS OBSERVED | | DECOMPOSITION OBSERVED | |
|---|---|---|---|---|
| | No insects | Insects | No insects | Insects |
| 1 | 0 | 5 beetles<br>2 flies | Turning brown<br>Smell moderate | Meat liquifying<br>Smell moderate |
| 2 | 0 | | | |
| 3 | 0 | | | |
| 4 | 0 | | | |
| 5 | 0 | | | |
| 6 | 0 | | | |
| 7 | 0 | | | |
| 8 | 0 | | | |
| At conclusion | | | | |

## Research Suggestions

- Study how termites decompose wood. How might the termites' symbiotic relationship with organisms in their gut be affected by environmental stress such as acid rain?
- Read more about scavengers, decomposition, and the recycling of organic nutrients.

# 15

# Aquatic insects
## Gone fishing

*(Aquatic-insect habitats and microhabitats)*

About 90% of all insects are terrestrial—but about 10% are aquatic, living in fresh waters. Insects are important members of aquatic ecosystems primarily because they are food for fish, waterfowl, and other organisms in these ecosystems.

Freshwater habitats are categorized as either *lotic* ("running") or *lentic* ("standing"). Organisms found in these two types of habitats have different physical needs. Organisms found in bodies of running water must have some method of attaching themselves to the substrate if they are not to be washed downstream. (See FIG. 15-1.) Free-floating plankton is common in standing bodies of water, but not in running waters. Both lentic and lotic habitats have smaller microhabitats existing within each. For example, a stream may have eddies in some parts and riffles in others, both with their own characteristics.

## Project Overview

Fly fishermen use aquatic insects, both adult and larval forms, as models for artificial bait. (See FIG. 15-2.) They tie and use different flies depending on the different types of fish they are trying to catch as well as where and when they are fishing. As mentioned above, there are reasons why insect populations differ in running and standing bodies of water. This project will not only investigate the different insects living in lentic versus lotic waters but also the different microhabitats existing in the same body of water. Are there different types of insects found within the same

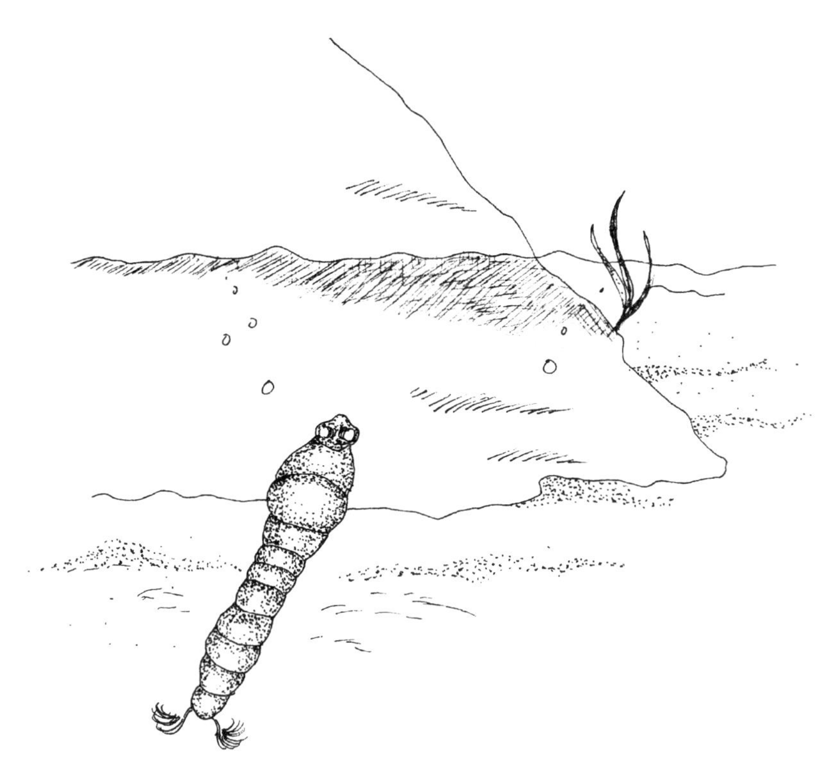

**Fig. 15-1** *Many aquatic insects attach themselves to substrate such as this rock.*

**Fig. 15-2** *Fishermen create flys to mimic aquatic insects.*

river or stream? Is there scientific merit to a fly fisherman's using different flies within the same body of water? To answer these and any other questions your research leads you to, begin your literature search and formulate your hypotheses.

## Materials

- Access to a standing body of water (such as a pond or a lake) and to a running body of water (such as a stream or a brook)
- Aquatic collecting net (If this net cannot be purchased from a supply house, you can use a large, very fine sieve.)
- White pan or white bucket
- Forceps
- Heavy rubber gloves
- Rubbing alcohol (70%)
- At least 10 glass jars or vials with screw caps that can be closed tightly
- Labels to apply to the jars
- Magnifying glass (optional)
- Plastic grocery bags
- Strong string
- Field guide that specializes in aquatic insects (These guides usually concentrate on immature forms. See the reference section at the end of this book.)

## Procedures

In the first part of this experiment, you will survey the insects living in a standing body of water versus a running body of water. In the second part, you will collect specimens from various microhabitats within the running body of water.

Before you start collecting for the first part, prepare the jars by filling them about halfway with rubbing alcohol, which will preserve the insects. Label the jars "pond/vegetation," "pond/bottom," "stream/vegetation," "stream/beneath rocks," and so forth. You can omit the second word of each pair until you actually collect the specimens. Use a lead pencil to write on the labels since ink will blur from water or will be dissolved by the alcohol if it spills.

CAUTION: Whenever working in or near water, be sure that another individual is with you and that you have approval from your sponsor.

Now you can begin collecting. While standing on the edge of a pond or lake, sweep the net or sieve through the water, especially in areas where vegetation is growing out of the water. (See FIG. 15-3.) If there are different types of emergent plants, collect in each area where the different vegetation grows. After you've collected from each of these habitats, dump the contents of the net or sieve into the white pan. Using forceps,

**Fig. 15-3** *Collect insects by sweeping an aquatic net through the emergent vegetation.*

look through the debris for insects. As you find the insects, use the forceps to place them in the appropriately labelled jars. If you haven't already labeled the exact locations on the jars, do so now.

Next, collect in an area where the pond is very shallow by dragging the net or sieve across the bottom of the pond. Once again, dump the contents into the pan and place the insects into their proper jars. Finally, while wearing gloves, turn over some rocks near the edge of the water, and using your forceps, collect any insects clinging to the bottom and place them in their vial.

After collecting at the pond or lake, go to a stream or brook and collect from a number of different habitats. Dump your specimens into the pan and place them into properly labeled vials. First, collect from the surface by sweeping the net or sieve through the water, preferably where there is vegetation. Next, collect from the bottom of the water in a shallow area by scraping the bottom of the stream with the net or sieve.

After collecting in this manner, have your partner use a stick to disturb the bottom of the stream while you hold the net in the water downstream to catch insects (and debris) that has been dislodged by the stick. This is called kicknetting. Do this for 1 or 2 minutes. Dump the collection into the pan, locating and preserving the insects, as explained above. For the final collection, put on gloves, turn over some rocks near the edge of the water, and use your forceps to collect any insects attached to the bottom. Finally, dump this collection into the pan and locate and preserve this group also.

When you have completed the first part of this project and while you are still at the stream, you can begin the next part of the experiment—namely, to determine whether different types of insects inhabit different microhabitats within the stream, such as the riffle (or fast part of the stream).

Use plastic grocery bags to cut five pieces of plastic, 1 inch wide and about 2 feet long. Tie each piece around a fist-sized rock (about 1 foot of the plastic should be hanging from the rock). (See FIG. 15-4.) (You might also want to wrap string around the rock and around the tied portion of the ribbon to ensure that the ribbon won't come off). Do this for five rocks. Put the rocks in different parts of the stream, placing some in areas where there is almost no motion and others where there are varying water speeds. Some rocks can be simply placed along the shore in eddies, inlets, or in the rapidly running portions. Attach a long string to a few rocks and throw the rocks well into the main portion of the water, leaving the end of the string on shore where it can be retrieved.

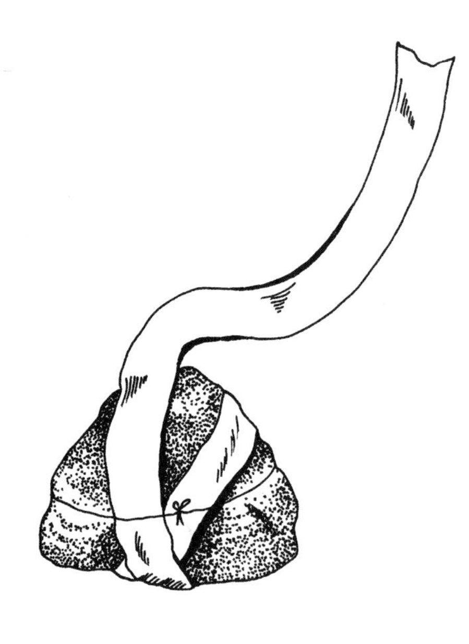

**Fig. 15-4** *Tie the plastic strip to a rock to collect underwater insects.*

Return the next day to be sure that the ribbons are still attached to the rocks and that they can be retrieved. Then leave the rocks in place for about 3 or 4 weeks, after which time you should return to collect them. Pull the strings in or, while wearing gloves, simply pick them up if they are just offshore. Observe how many insects are attached to or crawling on each plastic strip. Use your forceps to pick off these insects and place them into the appropriate vials.

After collecting the insects you can begin to identify them. Use a magnifying glass or a dissecting microscope, if available. Draw sketches of each organism collected. Then note how many of each were found. Use your aquatic-insect field guide to determine what kinds of insects you found in each habitat. Many of the insects that were found will be immature forms.

## Analysis

What kinds of insects did you find in the lentic and the lotic habitats? Were the same insects collected in both? If the insects collected in either habitat differed, how did they differ structurally? If the ecology of the stream differs enough from the ecology of the pond to cause different insects to inhabit each, what factors are responsible? (Some of the differences will not be obvious from your collections but should become obvious from your literature search.)

Analyze the insects that were collected from the different parts of the stream in part two of this project. Which rocks had the most—and which the least—number of insects? How important are the microhabitats to the insects inhabiting them? Are the differences as great within a habitat as between habitats? If differences exist, what factors make them different? Is there a scientific basis for using different flies in different parts of a river or stream?

## Going Further

- Speak with experienced fly fisherman about their favorite flies. Investigate what insect the fly mimics. Can you determine whether the fly attracts fish as well as the actual insect does?
- Can you tell if a certain fly moves through the water in a manner similar to the actual insect that it mimics?

## Suggested Research

- Read more about the life cycles of aquatic insects. Do these insects spend most of their lives in the water? If not, how much of their lives are spent out of water? What do most of these aquatic insects feed upon?

PART V

# Insect Behavior

In the simplest sense, behavior is defined as what an animal does. The biological study of animal behavior is called ethology. There are many types of behavior including feeding behavior, defensive behavior, grooming behavior, social behavior and reproductive behavior. Most insect behavior is innate—insects are born to behave in a certain way. Innate behavior can be as simple as the reflexive extension of a fly's mouthparts when it senses food or as complex as the way in which an insect orients itself in flight.

Some insect behavior, however, is learned. Unlike innate or instinctual behavior, which involves using information that one is born with, learned behavior comes from experience and involves accumulating information during one's lifetime. Honeybees, for instance, can learn the way to get back to their hive. In many instances, innate behavior can be modified by learned behavior, so that they are often associated with one another. Both innate and learned behavior helps the insect to survive—by choosing the correct food, locating the correct kind of shelter, or even finding the right mate. Behavior also includes communication and social intercourse between insects. Some insects can communicate with each other by using chemicals, songs, or body language—all of which are aspects of behavior.

The most important part of any behavioral study is to accurately observe and describe how an insect behaves. This accurate description of an insect's behavior is called an ethogram. To study behavior you must chart the movements of an individual before, during, and after some event. The event could be the change of light to dark; a change in temperature, of habitat, of food, of neighbors; or a physical or chemical disturbance called a stimulus.

When you study behavior and report the results, it is important to be completely objective. Don't make assumptions about the reason for the behavior; just report the facts. Try not to draw conclusions from the facts. It's easy to say, "The insect was unhappy," or "The insect loves the light," but these are assumptions. Students have a tendency to apply anthropocentric observations to behavioral studies—that is, they tend to apply human values to other organisms. For interesting reading about insect behavior read Konrad Lorenz's *King Solomon's Ring*.

# 16

# Taste

## One man's ceiling is another man's floor

*(Taste receptors in blow flies)*

Sensory perception is a fascinating aspect of biology. While most of the higher animals have rather straightforward sensory organs, many of the lower animals have unique ways of sensing their environment. For example, when you think of taste, you usually think of a tongue; but many animals don't have tongues, yet must still be able to perceive taste. Lower forms of life use a multitude of organs to sense "taste."

## Project Overview

Insects sense their environment in many ways, including with their eyes and antennae. But these are not the only sensory organs used by insects. How do some insects taste if they don't have a tongue? By careful observation, can you find other sensory structures on insects? The blow fly is a good example of an insect that uses unusual parts of its body to sense the environment. How does a blow fly "taste" its food?

When many types of flies taste food, they extend a *proboscis* (or tube) down to the food source. (See FIG. 16-1.) With this behavior in mind, what organ (or organs) does the blow fly use to "taste" its food? To answer this question and any others that your research leads you to, begin your literature search and formulate your hypotheses.

## Materials

- At least two live adult blow flies, (These can be purchased from a supply house and usually come in the pupal form, with adults emerging within a week.)
- A refrigerator
- Rubber cement
- Applicator sticks (such as wooden coffee stirrers or popsicle sticks.)
- Forceps or tweezers
- Cup
- Teaspoon
- Two saucers
- Magnifying lens or dissecting microscope

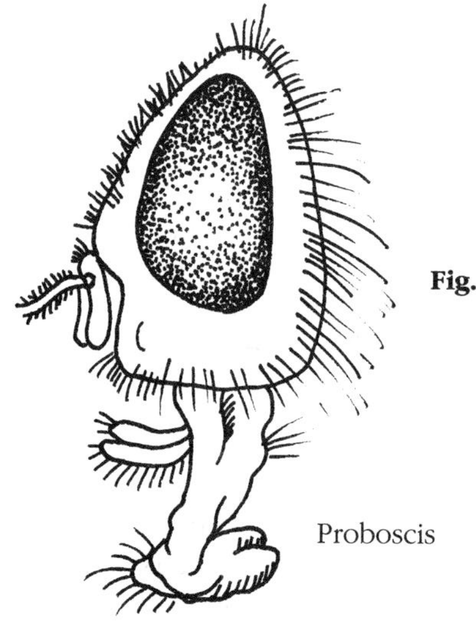


**Fig. 16-1** *Many flies lower their proboscis when feeding.*

## Procedures

Before beginning this experiment, it would be beneficial to have an illustration of the external anatomy of a blow fly (or similar fly). Such an illustration would help you with your data collection and your own drawings.

You will create a setup similar to the one in FIG. 16-2. First, the flies must be immobilized without harming them. To slow the flies down, place the container of flies in a refrigerator for about 10 minutes. When the flies are no longer moving, open the container and with forceps gently take out one of the flies. (Put the other flies back in the refrigerator.) Place a small drop of rubber cement directly on the back (top) of the fly, between its wings. (See FIG. 16-3.) Place another drop of glue at the end

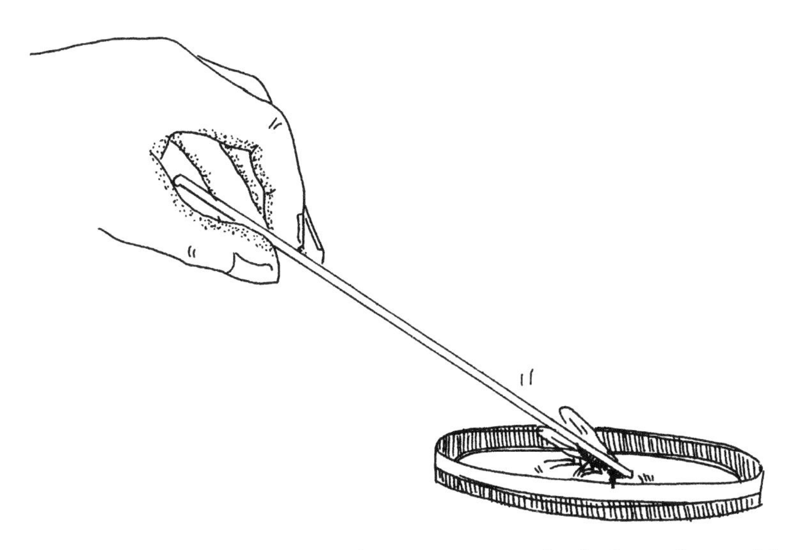

**Fig. 16-2** *The fly is attached to a stick so various parts of its body can be tested for "taste" reception.*

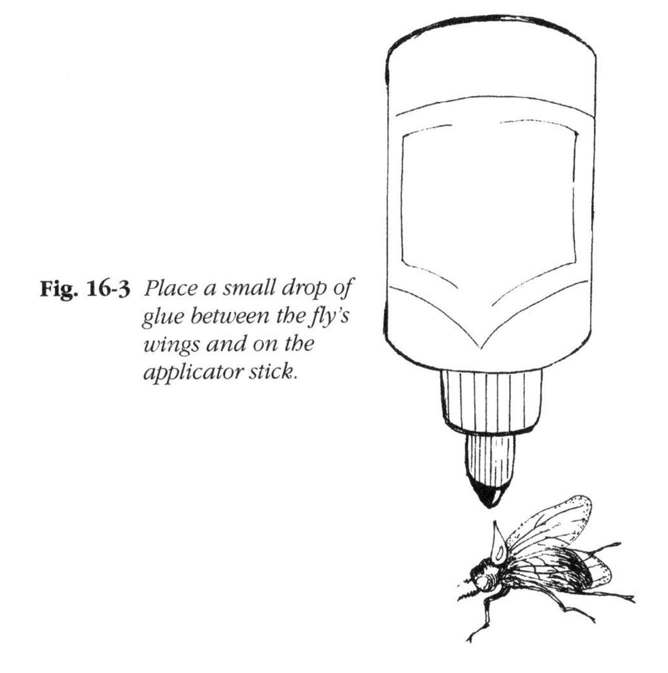

**Fig. 16-3** *Place a small drop of glue between the fly's wings and on the applicator stick.*

of an applicator stick and touch the two globs of glue together, so that the fly is attached to the stick and hanging down as shown in FIG. 16-3.

Set the stick, with its attached fly, in an upright position in the cup so that the fly isn't harmed. Repeat this process with the other fly and place it in the cup, too. After a few minutes, the flies will recover from the cold and will try to fly—but will be stuck to the sticks.

The flies should not be thirsty (just hungry), so let them drink water before you begin the experiment. To do this, pick up one of the sticks and allow the fly's mouthparts to barely touch the water. You'll notice its proboscis lapping up the water. Let it drink as much as it wants. When it is no longer drinking, put the stick back in the cup and do the same with the other fly. Leave both flies in the cup for 10 minutes and let them settle down.

Now find out what body parts the blow fly uses to taste food. Create a sugar solution by mixing in a saucer 3 teaspoons of sugar in 4 tablespoons of water. Hold one of the sticks with the attached fly and carefully dip different parts of the fly's body into the sugar-water solution. First, just barely touch the tips of the fly's legs into the water. Does its proboscis extend to feed? Now try the top of its head above its eyes. Did the proboscis extend? Try the wings and try the tip of its abdomen. How about the proboscis itself? Does it extend when it touches the water? To confirm your results, repeat the entire procedure with the second fly.

## Analysis

From your observations, can you tell which parts of the fly's body sense taste? Fill in a table similar to TABLE 16-1. Can the fly taste with more than one part of its body? Use a dissecting microscope to look at the structures that you have determined are used for taste. What portion of these structures do you think is actually responsible for sensory perception? If more than one part of the insect was found to elicit a response, look at the

**Table 16-1**

| INSECT STRUCTURE | PROBOSCIS EXTENSION Yes or No | |
|---|---|---|
| | FLY #1 | FLY #2 |
| Mouthparts | | |
| Tarsi on forelegs | | |
| " " mid legs | | |
| " " hindlegs | | |
| Head | | |
| Wingtips | | |
| | | |

parts under a dissecting microscope. Do the parts have structures in common that might be used for sensory perception?

## Going Further

- To continue this experiment, test to see the minimum concentration of sugar that elicits a response. Make up various sugar-water concentrations. Dip the fly (on the stick) into different concentrations and make notes on the movements of the mouthparts. Does the fly taste all of the concentrations or only a few? What is the minimum threshold concentration that elicits a response?
- Devise an experiment to determine if the anatomical structures identified in this project are actually responsible for sensory perception.

## Suggested Research

Read more about the sensory structures of insects. How do they taste, hear, touch, smell? Do all insects sense their environment in the same way?

# 17

# Pheromones

## Abandon ship!

*(Alarm pheromones using ants and aphids)*

Pheromones are chemicals secreted by some insects to communicate information to other related insects. These chemicals can be thought of as external "hormones" whose targets are not organs (as with normal hormones) but other individuals. Pheromones are often used to transfer information among members of an insect colony, including things such as who is sick, hungry, or in danger. Many ants and aphids use alarm or warning pheromones to broadcast impending danger.

## Project Overview

*Aphids*, also called plant lice, are common insects. They live in colonies on plant stems and feed by sucking the sap. (See FIG. 17-1.) Aphids insert their mouthparts into the plant, where they remain indefinitely. If undisturbed, an aphid can spend its entire life in the same spot where it was born. This does not usually occur, however, since aphids are a favorite food for many insect predators.

Predators like the ladybird beetle enjoy marching and munching through colonies of aphids. The immature form of the ladybird beetle is especially voracious and capable of devouring enormous quantities of aphids.

When an aphid is in danger, it releases an alarm pheromone. If an aphid is caught by a predator, it secretes this pheromone—thus alerting other aphids to the danger. Aphids can't fight, so their only available defense is to flee.

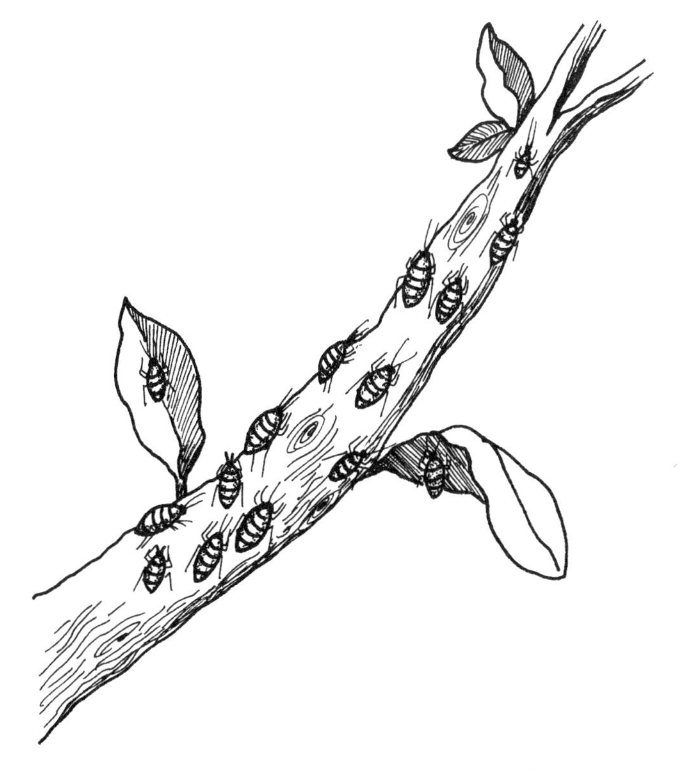

**Fig. 17-1** *Colonies of aphids are easily found in warm weather in most parts of the country.*

Ants also use pheromones to communicate danger. Like an aphid, when an ant is injured it releases a pheromone to warn others, but the response is different from that of the aphids.

In this experiment you will use both aphids and ants (if available) to determine how quickly the pheromones work, how far away individuals can be and still sense pheromones released by others, and how other individuals respond to pheromones. Do the pheromones work instantaneously? What is their range? What is the behavioral response of other individuals to the pheromones? Do all individuals react in the same way? How long do the pheromones last? To answer these and any other questions your research leads you to, begin your literature search and formulate your hypotheses.

CAUTION: Ants can bite. Follow the instructions carefully. There is no need to come in direct contact with ants in this experiment. Some portions of the United States contain *fire ants*, which are dangerous and must be avoided. If you live in an area where fire ants are present, don't perform this experiment unless you can identify and avoid these ants.

## Materials

- An aphid colony covering at least 3 inches of a plant stem (You can find a colony in a local field or garden, or you can create your own by placing a bean plant outdoors in a very sunny spot during the summer months in most parts of the country. It will be naturally infested with aphids within a few weeks. During colder months it might be possible to find aphids in a local greenhouse.)
- Ant colony (The easiest way to perform this experiment is to find an active colony of ants outdoors. If you prefer, you can buy a complete kit from a toy or science store, or you can build your own ant colony container by following the instructions at the end of this project. If you're building your own ant colony, live ants can be purchased from a scientific supply house, or you can collect your own ants by following the instructions at the end of this project.)
- A couple of ants of a different species or from a different colony (You can either collect your own by following the instructions at the end of this project, or you can purchase another type of ant.)
- Fine forceps
- Quality ruler
- Stopwatch

(If you plan to build your own ant colony chamber you'll need additional materials. See the section at the end of this project.)

## Procedures

This experiment is best done from spring to fall, when aphids and ants are easily found. The project consists of two parts (aphids and ants), which can be performed in either order. In both experiments, you will apply a stimulus that simulates a predatory attack on an individual member of the colony. You will record your observations before and after the stimulus. Each stimulus will be applied multiple times until you collect all your data. Documenting your observations of insect behavior is crucial to the success of this project.

### Aphids and pheromones

Locate or create an aphid colony as described in the "Materials" section. Observe individuals in the aphid colony for a few minutes. Watch their movements (if any). Do not touch or disturb the colony in any way. Record your observations.

At the center of the aphid colony, midway on the stem, use a fine forceps to gently squeeze a single aphid. (See FIG. 17-2.) This stimulus mimics the effect of a predator, such as another insect. Be sure not to disturb any other aphids as you do this. Apply the forceps to the front por-

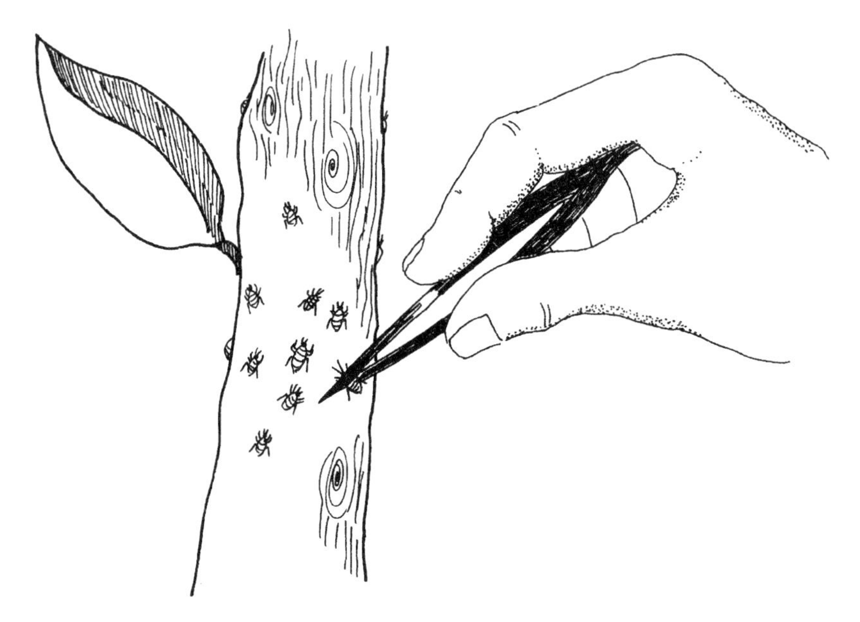

**Fig. 17-2** *Gently squeeze the anterior portion of an aphid, but don't knock it off the stem.*

tion of the insect's body. Release the forceps (the injured aphid must remain on the plant). Start (or have someone start for you) the stopwatch to time the events. Quickly and carefully observe the behavior of the other aphids on the stem. Write down your observations. Check the watch to see how long it took for other aphids to respond and how long their response lasted. Begin by concentrating your attention on those aphids immediately adjacent to the injured aphid. Repeat the procedure a few more times to confirm the results.

Now, if possible, collect an aphid from another area. Try to find an aphid of another species. Use your insect guide to identify the different species of aphids. Place the "foreign" aphid on the stem in the midst of the other indigenous individuals and perform the same experiment. You can also try this with members of the same species, but from a different plant found in a different location. Once again observe and document what you see.

## Ants and pheromones

Buying or creating your own colony means that you can perform this part of the experiment at any time of year. If you plan to find a colony outdoors, however, the experiment must be done during warm weather.

Find an active colony of ants outdoors or create your own colony as mentioned in the "Materials" section. Begin by watching the normal behavior of individuals in the colony and recording your observations.

Take your time and observe as much detail as possible. What do the individual insects do? What body movements do they make and where do the insects go? How do the ants behave with one another? Do they touch each other or avoid each other?

If you are using a colony that has been purchased, record the behavior of the colony when the lid or top of the colony chamber is opened and closed. This will assure that the observations you are about to make aren't due to the opening and closing of the top but to the test stimulus.

Once you have finished observing this "normal" behavior, use your fine forceps to squeeze a single ant in the colony. (See FIG. 17-3.) Do this in a location near a lot of other ants. In an artificial chamber, lift the top off the colony, reach in and with the forceps squeeze an ant on the surface of the sand. If it's a natural colony, squeeze an ant on the soil near the opening to the nest. Use the stopwatch to time the responses of the other ants—when the responses start and how long they last. Document your observations.

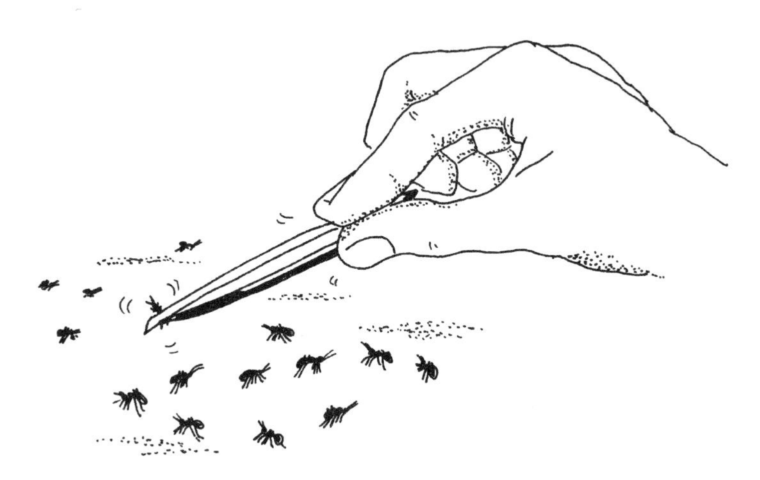

**Fig. 17-3** *Gently squeeze an ant and notice the response of other individuals nearby.*

Observe those individuals near the injured ant and those farther away. Take measurements with the ruler. Did the behavior of the other ants change? Continue your observations until the colony's behavior appears to be normal. Repeat this procedure a few times in different areas until you have confirmed the results and feel confident about your data.

When this is completed, find an ant (or use the one you've purchased) of a different species, or at least from a different colony. Place this "foreign" ant into the colony and crush it with your forceps in a similar fashion. Perform this portion of the experiment in the same way that you did the first portion. Record your detailed observations. Repeat this procedure to confirm the results.

## Analysis

First analyze each experiment independently and then compare your findings from both experiments. How long did it take for other individuals in the colony to respond? How did the other individuals respond? How far from the damaged insect did the response occur? How long did the pheromone appear to have an effect on other members of the colony? What happened when the "foreign" insect was introduced? Were the results the same for both types of insects? How did the responses differ?

## Going Further

Concentrate your study on the different responses that you observed when members of a different colony (but similar species) and members of a different species were introduced. How did the response differ when members of the same species were used? Just how specific are pheromones?

## Research Suggestions

- Investigate how pheromones are used for pest control.
- Read more about the social behavior of aphids and ants and the role that pheromones play.
- Research whether any human pheromones have been found. (See FIG. 17-4.) Read the scientific literature and also read advertising and marketing claims from companies that sell perfume.

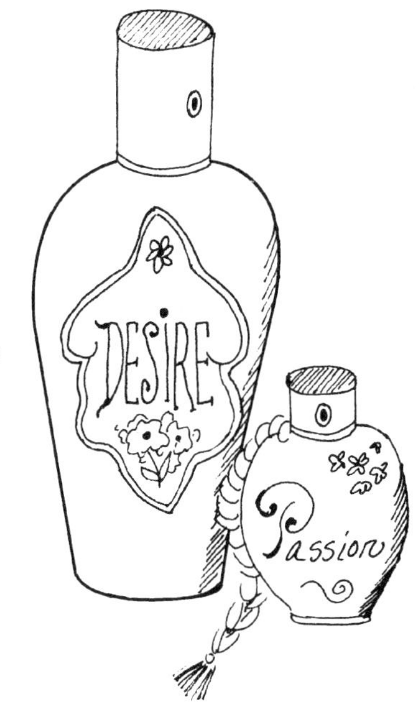


**Fig. 17-4** *Have human pheromones been discovered?*

## Building Your Own Ant-Colony Chamber

If you plan to build your own ant-colony chamber "from scratch," you'll need the following additional materials.

- A large, clear jar with a wide mouth (e.g., 32-ounce mayonnaise jar)
- A smaller-mouthed jar (e.g., smaller mayonnaise jar ) that can fit inside the larger jar and leave a ½- to 1-inch gap between the two jars
- Sand
- Sponge
- Honey
- Fine nylon netting (such as pantyhose)
- Rubber bands
- Small package of seeds (any kind)

Turn the smaller jar upside down and place it in the larger jar. Pour the sand in the space between the two jars until it covers the inside jar. (See FIG. 17-5.) Sprinkle the seeds on the sand's surface and then place five drops of honey on the surface of the sand for food. Cut a 1-inch-square piece of sponge, dampen it, and place it on the surface of the sand. This sponge will be the ants' water supply. Cut a double or triple layer of the fine nylon netting so that it can be used as a cover over the (larger) jar mouth, and use rubber bands to hold the netting in place over the jar's mouth. Be sure that the cover is secure.

**Fig. 17-5** *Create your own ant colony chamber by pouring the sand over the inner bottle to fill the gap.*

## Collecting Your Own Ants for the Colony

To collect your own ants you'll need the following additional materials.

- Empty tuna can
- Honey

- Forceps for holding the can
- Large, heavy-duty, zipper, plastic baggy for holding the can (with ants)
- A closable container that can hold the baggy, the can, and the ants

During the warm weather, collect ants by putting a teaspoon of honey in an empty tuna fish can and placing it outside in an area where you have noticed ants. When a large number of ants are in the can, use the forceps to pick it up and put it quickly in a sealable plastic bag.

Place the baggy, can, and ants into a closable container, such as a rubber storage container (be sure it is tightly sealed.) Put the container (with everything in it) in the refrigerator for at least 15 minutes to slow down the ants. When the ants appear to be immobilized, shake them into the ant colony. Place the nylon cover over the jar and secure with rubber bands (be sure the jar is tightly sealed). Keep the colony in a safe place, but not in your home, in case the container breaks or the ants otherwise escape.

## Collecting a "Foreign" Ant

Search for ants in an area at least 100 yards away from where you collected the original colony of ants. (If you purchased the colony, it won't matter where you find the foreign ant.) When you locate the ant, either slide a piece of paper underneath it, pick it up, and place it in a small vial, or use a transfer aspirator.

# 18

# Protective coloration

## Nature plays hide & seek

*(Does protective coloration really work and is it found in many habitats?)*

Organisms have evolved a multitude of methods to defend themselves from predators. One of the simplest methods is protective coloration. Animals that cannot be easily seen are less likely to be located and devoured. Some animals even change colors to adapt to a new environment. Whether a color is protective or not depends on the organism's environment and the predator's visual capabilities.

Insects play vital roles in almost all ecosystems and are prey for numerous types of organisms, such as spiders, other insects, birds, rodents, bats, bears, fish, and even humans in some parts of the world. Insects use many methods to defend themselves against these predators. Flying, responding rapidly, protecting themselves with natural armor, using repellent chemicals, and faking death are a few ways that insects protect themselves.

## Project Overview

Many insects defend themselves by using protective coloration and protective resemblance. An extreme example of this is the insect called the walking stick that looks and even sways in a fashion similar to a small branch on a plant. (See FIG. 18-1.) Less dramatic are insects that simply match the color of their habitat, blending in with their background.

This project involves three parts. The first two parts concentrate on the predators' point of view regarding protective coloration, and the third part studies the variety of protective coloration in different habitats.

**Fig. 18-1** *The walking stick is a good example of protective coloration and resemblance.*

In the first part you'll examine whether protective coloration makes it more difficult for people to find an object. You'll do this by searching for multicolored macaroni on a plot of land.

Part two examines whether protective coloration makes it difficult for birds—the natural predators of insects—to find objects. Do birds, as well as other predators, find it more difficult to pursue insects that blend in with their background? To simulate insects, you'll use colored pieces of bread spread out on a plot of land.

The third part investigates how widespread protective coloration is in a variety of habitats, such as a green lawn, a brown field (of dead grass), the surface of soil, and leaf litter. What percentage of insects in a particular habitat blend in with their background? Is protective coloration common in the insect world? What colors are common in each habitat? If an insect lives on or in the soil, is protective coloration used, and does it work? To answer these and any other questions that your research leads you to, begin your literature search and formulate your hypotheses.

## Materials

- Dry, tricolored macaroni to simulate insects of different colors for the first part of the experiment (One of the colors must be green or brown to match the color of the plot of land you are using. You can purchase colored macaroni at the grocery store, or you can make your own by cooking the macaroni according to the directions on the box and by adding food coloring while cooking. If you prepare your own, make some green, brown, and a few bright colors like red, blue, and orange. Air-dry the macaroni.)

- White bread and food coloring for the second part of the experiment (Break white bread into small pieces and soak the pieces in different food coloring. Make the same number of each color. Let them air-dry. Mix all the colored pieces together. Make one of the colors green or brown to match the plot of land.)
- A lawn or meadow with about 2 inches of growth (preferably with consistently colored vegetation, such as all green or all brown)
- Lunch bag
- Yardstick
- String
- Watch with a second hand
- Insect-collecting net

- Killing jar and activating fluid
- Large white piece of paper or a white pan
- Tweezers or forceps

## Procedures

The first two parts of this experiment can be done at any time of the year, as long as the lawn or meadow isn't under snow. The third part (collecting) is best done during warm weather. You can perform only portions of this project, if you prefer.

The first part of the experiment examines whether (simulated) protective coloration works when you and a partner act as predators. Count out 20 pieces of macaroni of each color (as described in the "Materials" section). Put the pieces in the lunch bag and shake well.

Measure a plot of land 6 feet square and mark its boundary with string (simply lay the string on the ground around the perimeter). Select a plot that has vegetation consistent in color. Have another person randomly spread the macaroni within this area. With the other person timing you, pick up as many pieces of macaroni as you can in 30 seconds. (See FIG. 18-2.) At the end of this time, count the number of each color of the macaroni you picked up. Repeat this procedure three times and then reverse roles and have your assistant perform the experiment while you do the timing. Do this at least three times each.

Collect all your data. Use the average of all the runs. Was one color collected more frequently than others? What was the relation between the most commonly collected macaroni and the color of the field or the meadow?

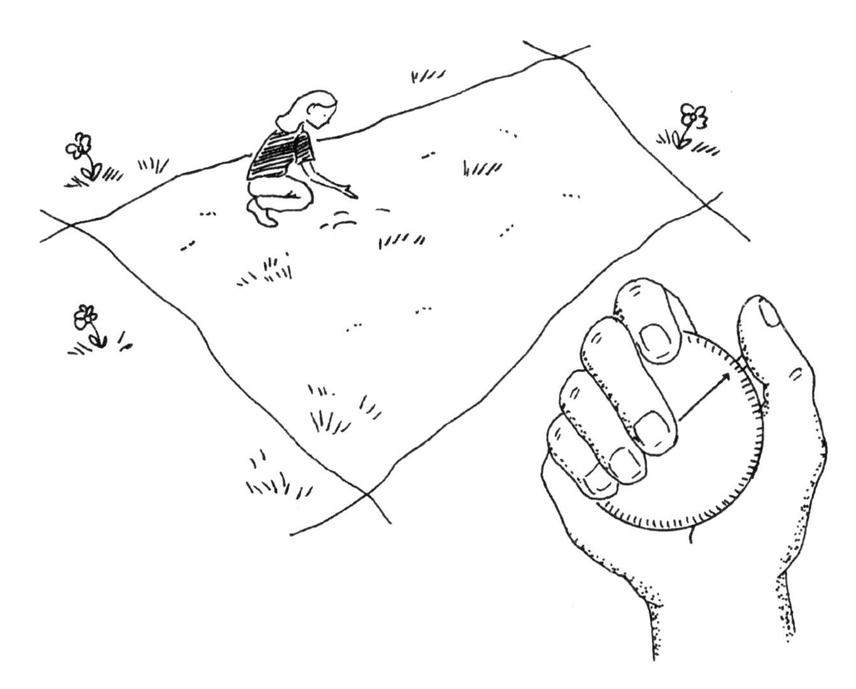

**Fig. 18-2** *Pick up as many pieces of tricolored macaroni in 30 seconds as possible.*

In the second part of the experiment, you'll examine whether (simulated) protective coloration has any value when a natural enemy—the bird—is the predator. Take an equal number of each of the colored pieces of bread (as mentioned in the "Materials" section) and randomly spread them out on your lawn, a field, or a meadow where birds are often seen. The vegetation should be very short and of one consistent color. Randomly spread the colored pieces of bread over the plot within a confined area as you did in the original experiment.

Once the bread is distributed, leave the area and observe from a distance (binoculars would be helpful). Wait until some birds land and begin feeding on the bread. Let the birds feed until about half the bread is gone. (See FIG. 18-3.) Go out and count how many pieces of each color were eaten. Collect your data. Repeat this procedure three times.

For the third part of the project, determine if protective coloration is truly a common phenomenon in the insect world. In this part of the experiment, you will collect insects from different habitats and categorize them by color to determine what percentage of the total number collected matched their habitat. For the first habitat, find a field or meadow that contains vegetation of one consistent color. The vegetation should be at least ankle high and preferably up to your knees.

**Fig. 18-3** *Observe the birds as they eat the colored pieces of bread.*

(Dress accordingly to protect yourself from bites, especially in areas where Lyme disease is carried by ticks. And check yourself for ticks afterward.)

While in the field, prepare the killing jar as described in chapter 1 of this book. Sweep net the field or meadow to collect insects living in the vegetation, as described also in chapter 1.

When you've collected and immobilized the insects, dump them out of the net and into the killing jar, and leave them there for about one hour. Pour the collection on white paper or into a white pan. Separate the insects by color (a fine brush or forceps will help manipulate the insects as you count, and a magnifying glass might also be useful). Count each color group. Then determine the percentage of the total for each color. Is one color more common than others? How many colors were found? Did some contrast with their background instead of blending in?

For the next habitat, locate an area with fertile soil, such as a garden, or an area with a lot of leaf litter, such as in the woods. Using forceps, pick out any insects or other arthropods that you can find on the surface of the soil or leaf litter and place them in the killing jar. After about one hour, pour them onto white paper or a white pan and separate them by color. Count each color group. Is one color more common than others? What percentage of the total matched their environment?

For the final habitat, search for insects (both adult and larval stages) in different areas by digging into the soil a few inches. Do this in a wooded area and around the foundations of buildings. Use forceps to pick up and collect what you find. Count the color of these specimens and analyze the percentage of each color.

Note: If you plan to create an insect collection (as mentioned in the introduction) as part of your project, save the insects for mounting and preservation.

## Analysis

From the first two parts of this experiment and your literature search can you conclude that predators (humans and birds) find protective coloration a deterrent? In the third part of the project, what percentage of insects collected from each habitat appeared to use protective coloration? Record your data in a table similar to TABLE 18-1. Did protective coloration appear beneficial in all the habitats tested? If not, why not? How common is protective coloration in the world of insects?

**Table 18-1**

| Habitat | Number with protective coloring | Total collected | % of total with protective coloring |
|---|---|---|---|
| 1 | 28 | 36 | 80% |
| 2 | 1 | 18 | 5.5% |
| 3 | | | |
| 4 | | | |

## Going Further

- Devise an experiment to examine why some insects—such as some butterflies, beetles, bees, and wasps—are brightly colored. Is this a form of protective coloration or some other adaptation for survival?
- If insects from any of the habitats studied did not appear to use protective coloration, continue this study by rearing the insects through their entire life cycle. For example, if you found grubs (beetle larvae) that did not use protective coloration, does the mature form of the insect use it? What are the advantages for an insect that can change form throughout its life?

## Suggested Research

- Investigate other forms of defense used by insects. Is protective coloration the most common?
- Read more about insect defense mechanisms.
- Read about various forms of mimicry.

# 19

# Chemical communication

## Happy trails

*(Chemical communication among ants)*

Insects that live together in large nests or hives are called *social* insects. Termites, honeybees, and ants are social insects that live in nests or hives and cooperate with other members of their colonies in all aspects of their lives, including food gathering, defense and rearing the young. This cooperation depends on their ability to communicate with each other. People use spoken language or sign language to communicate with one another; but, obviously, insects can't speak and don't sign. Instead, they use other forms of communication, including pheromones and other chemicals. Insects are often highly dependent on these chemicals for survival.

## Project Overview

Ants use chemicals to communicate with one another. Ants will wander for long periods of time in search of food. When one ant finds a food source, it will release a chemical trail on the ground to lead it and the rest of the colony back to the food. The other members of the colony simply follow this trail to the food. In the first part of this experiment, you'll study what happens when the chemical trail is disrupted. How much of a break can occur in the trail before the ants can no longer find their way? What behavior patterns do they display when the trail disappears in front of them?

When this behavior is understood, you'll determine how much information this trail provides to the ants. Does the chemical trail that an ant creates simply show the direction of the food from the nest, or does it also indicate the route to the food? Is the information contained in the chemical trail specific to one species of ant? To answer these and any other questions that your research leads you to, begin your literature search and formulate your hypotheses.

CAUTION: Many ants can bite. Follow the instructions care fully. There is no need to come in direct contact with ants in this experiment. Some portions of the United States contain *fire ants*, which are very dangerous and must be avoided. If you live in an area where fire ants are present, don't perform this experiment unless you can identify and avoid them.

## Materials

- Five pieces of flat posterboard (about 12 inches square)
- Honey
- At least three (preferably five) active colonies of different species of ants that you've located outdoors
- A stopwatch or watch with a second hand
- Spoon
- Scissors or scalpel
- Work gloves

## Procedures

This experiment must be done in the late spring to fall, when ant colonies can be found. Prepare four of the pieces of posterboard by scoring strips about 1 inch in width. (See FIG. 19-1.) Score them deeply enough to make them easy to rip off (you can rip one of the strips off for a test run).

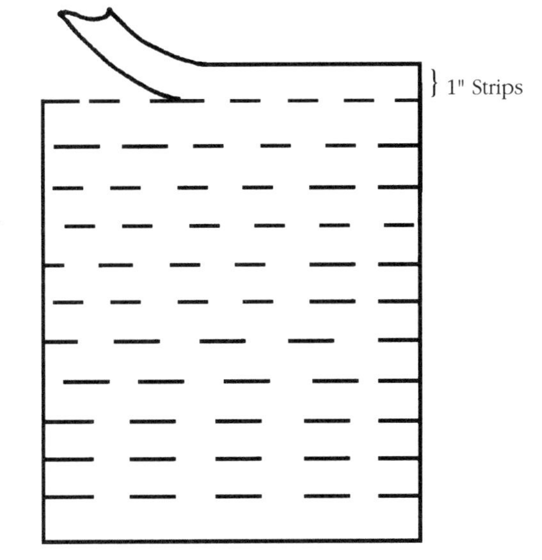


**Fig. 19-1** *You must be able to easily rip off each strip of posterboard.*

When they are ready, locate a colony or a well-defined trail of active ants. Place a large spoonful of honey a few feet away from where the ants are active, preferably on a hard surface such as a sidewalk or relatively flat earth. Surround the honey on all sides with the four scored pieces of posterboard. (See FIG. 19-2.) The edges of the posterboard must be as flush with the earth as possible. You can cover the outside edges of the posterboard with dirt to ensure that the ants will walk over the posterboard instead of under it.

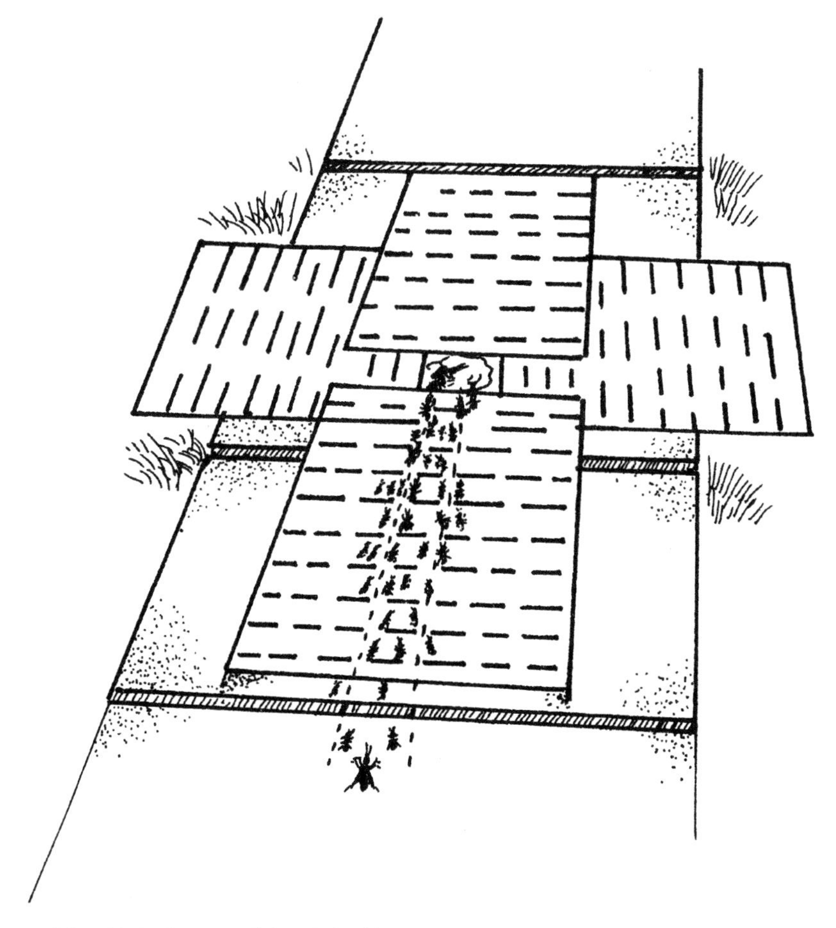

**Fig. 19-2** *Surround the glob of honey with four pieces of the posterboard.*

Soon, one or more of the ants will find the honey and lay down a chemical trail to it. After a while, other ants will pass over at least one of the pieces of posterboard as they form a trail leading to the honey. Record the behavior of the ants on the trail for at least 10 minutes. Look at the most minute movements that they make with their bodies and with their body parts, such as antennae. Use a magnifying glass and follow one individual. Record your observations.

Once the trail is established and your observations are complete, hold the posterboard (containing the ants) firmly down with one hand and rip off the first strip along the edge nearest the honey (a few ants will be on the strip, but place them out of the way). (See FIG. 19-3.) Observe the behavior of the ants as they come to the break in the trail. What do they do? Take detailed notes. Use a stopwatch to indicate time during your observations. How long does it take until the ants reestablish the trail?

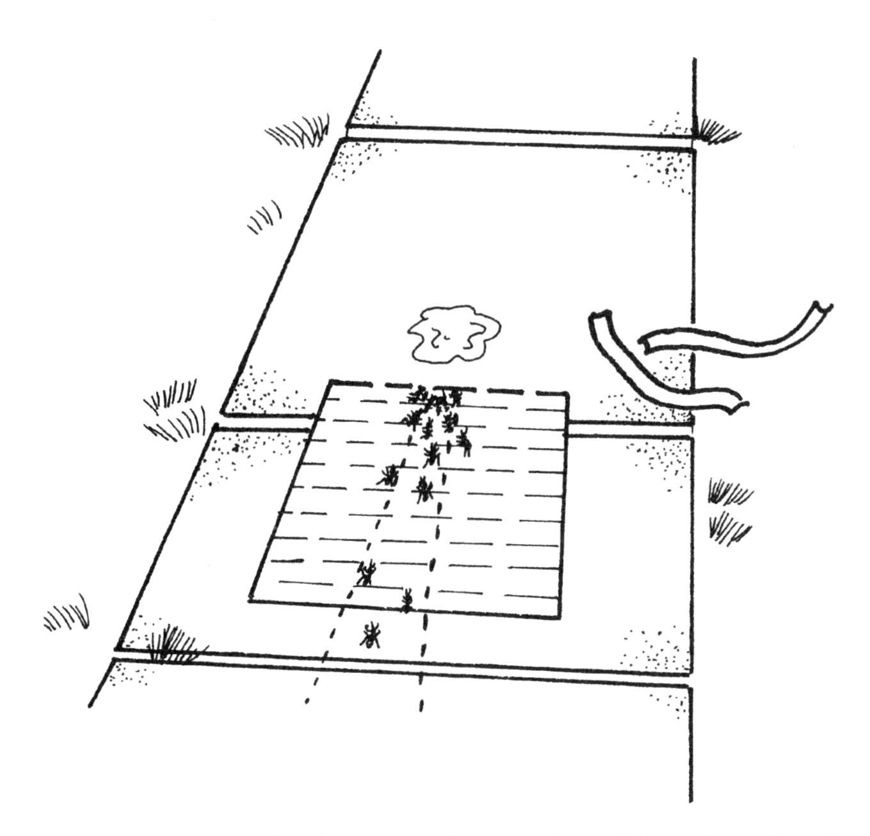

**Fig. 19-3** *After a strip has been removed, observe the ants' behavior as they approach the break.*

When the trail has been reestablished, rip off two 1-inch strips from the posterboard and make the same observations. How long does it take before the ants reestablish the trail? Are they always able to quickly reestablish the trail? Continue this procedure until the results become repetitive. You can now move on to the next part of the project to determine how much information is found in these trails.

Use the remaining piece of posterboard (this one was never scored) to replace the one that is now destroyed. Reestablish the trail once again as you did at the beginning of the first part of this project. When there are at least 30 ants coming and going over the posterboard, pick it up and turn it around so that the path is reversed 180 degrees. Do this quickly but gently so that you don't disturb the ants (wear protective gloves and don't let the ants crawl on you).

When you have made the switch, observe the ants' behavior. Note especially the ants as they leave the posterboard and as new ones enter it. What is their response? Compare it to the earlier behavior. Is there chaos, or do the ants simply continue on their way? If there is chaos, how long does it take for the ants to reorient themselves? Record your observations, including the time intervals. Repeat this experiment in three or four other areas, with a different ant species. Do you get the same results?

## Analysis

What were the behavior patterns that you observed when the trail was broken? Were they consistent? What happened when the break became longer and longer? At what distance did the ants become incapable of reestablishing the trail?

For the second part of the project, compare your data before and after changing the orientation of the posterboard. Can you conclude that the chemical trail indicates not only the direction but also the path to take? Were the results the same for all the species of ants? How important does this chemical-communication trail appear to be to these ants?

## Going Further

- Continue this experiment by picking up an established chemical trail (on the posterboard), knocking off all the ants, and placing the posterboard adjacent to a trail of ants from another colony. (See FIG. 19-4.) Do this first with another colony of the same species, then place the posterboard near the trail of a different species. Do the other ants follow the trail?
- Can you devise an experiment to determine if a chemical trail can be used to control ants in the home?

## Suggested Research

- Research the chemical composition of this trail. What is the active ingredient that the ants respond to?
- Research or perform experimentation to determine how the ants sense the chemical trail.

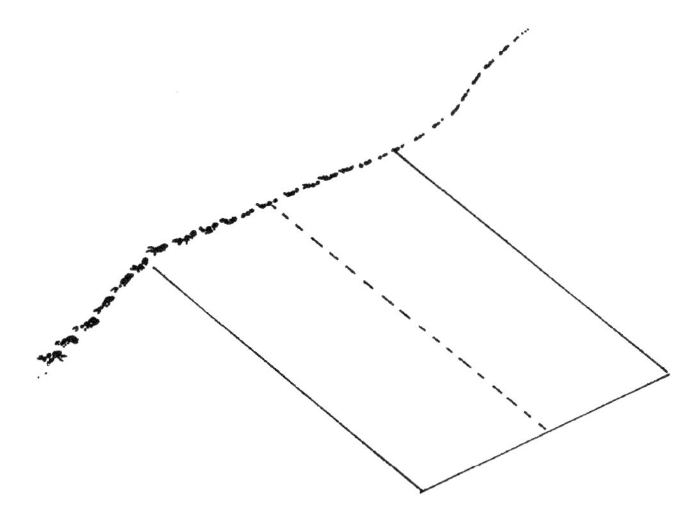

**Fig. 19-4** *Place the established trail perpendicular to an ant trail of another species.*

PART VI

# Domestications

Although all the projects in this book in some way can be linked to human lives, the following four projects are intimately related to our daily routines. *Organic gardening, organic farming* and *organic produce* are terms we are hearing more and more about each day. The term *organic* means different things to different people, but one of the primary factors that makes something organic is the absence of synthetic pesticides to control insect pests. The first project in this section investigates differences that exist between organically grown produce and regular supermarket fruits and vegetables.

Almost everyone uses insect repellents at one time or another, and there is no shortage of these products on the market. The next project studies how well the different repellents work (do some work better than others, if at all?) and whether health and beauty products, such as perfumes, negate the effect of these repellents? Another project looks at nature's replacement for processed sugar. Just how much sugar is there in a teaspoon of honey?

The final project investigates those insects that have moved in with us. With domesticated insects, our homes have become their homes. How have they adapted to our way of life? By studying how they live, might we learn ways to control them?

# 20

# Organic produce

## Beauty is only skin deep

*(How does insect damage differ when produce is grown organically and when it is grown with pesticides?)*

Even though synthetic pesticides and fertilizers are important in ensuring a substantial food supply, they pose many problems, including environmental dangers and health risks. Pesticide residues found on food have recently become a hotly debated issue. Laws govern the amount of pesticide residues that can exist on the food that we eat. The exact harm caused by eating many of these residues is unknown, but many have been linked to various cancers.

Pesticides also affect entire ecosystems through bioaccumulation and biological amplification. *Bioaccumulation* is the storing of pesticides in an individual's tissues over long periods of time. *Biological amplification* is the passing along of these pesticides to other organisms when that individual is eaten, resulting in a greater concentration of pesticides as they move through the entire food chain. Organisms at the top of the food chain, such as humans, end up with large quantities of these pesticides in their bodies.

Fertilizers seem harmless but cause environmental damage by a proccess called *nutrient enrichment.* Most fertilizers spread on farm fields and on homeowners' lawns never make it into the soil to reach the intended plants. Instead, they are washed away with runoff during rains. This nutrient-enriched runoff makes its way down gullies and streams and into ponds and lakes. The fertilizer acts like a vast reservoir of nutrients for algae, which grow with wild abandon and often choke the entire aquatic ecosystem. As the algae die, bacteria begin to feed on them, thus resulting in a bacterial population explosion. The large numbers of bacteria use up most of the available oxygen in the waters, causing other organisms to suffocate and resulting in a collapse of the aquatic ecosystem.

## Project Overview

The dangers of eating produce containing pesticide residues and the negative effect of these pesticides and fertilizers on the environment has made some people look for alternative ways to protect and provide for crops. Instead of buying fruits and vegetables grown on large, mechanized farms that use pesticides and fertilizers to excess, some people look to smaller organic farms that produce crops with natural pesticides and fertilizers.

When synthetic pesticides and fertilizers are not used, the crop yields are sometimes not as great and the prices are often higher. Organic produce doesn't contain pesticide residues, and the methods used in growing this produce have a far smaller negative effect on the environment.

Besides yield and price, however, what other problems might there be with organic produce? Is there more insect damage found on the food when these pesticides are not used? If so, does it in any way affect the taste or overall quality of the food? To answer these and any other questions that your research leads you to, begin your literature search and formulate your hypotheses.

## Materials

- A few apples and oranges from a fruit stand, store, or supermarket that sells organically grown produce
- A few apples and oranges from a regular supermarket
- Hand magnifying lens to observe insect pest damage
- A book that describes and illustrates apple and citrus fruit pest damage (This book describes the more common types of damage, but a more comprehensive manual would be useful.)

## Procedures

This project can be performed at any time of year, but you will obtain best results during the summer months when produce is abundant. There are two parts to this project. In the first part you will familiarize yourself with different types of damage caused by apple and citrus pests. (See FIGS. 20-1 and 20-2.) In the next part, you will survey the types and amount of damage found on these fruits in both a regular supermarket and an organic supermarket. You can assume that produce sold in a regular supermarket and not labelled "organically grown," has been treated with synthetic pesticides and fertilizers.

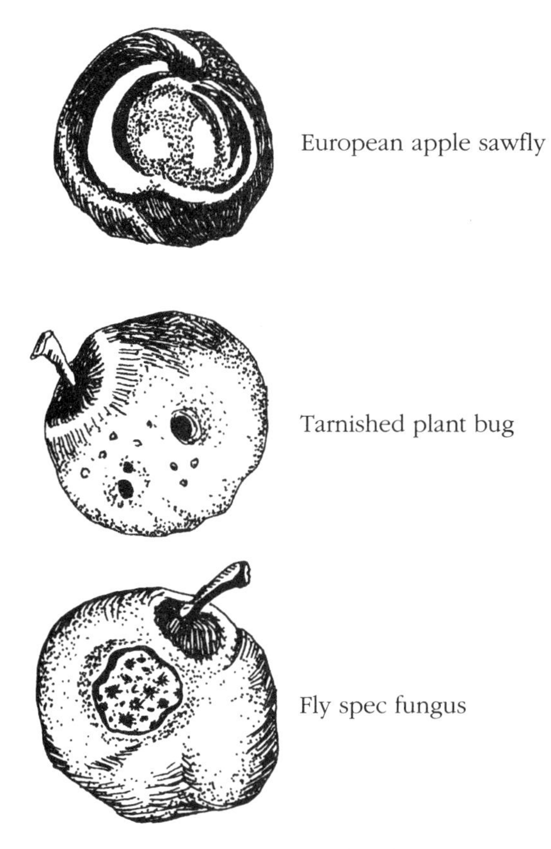


**Fig. 20-1** *Numerous types of damage are often found on apples.*

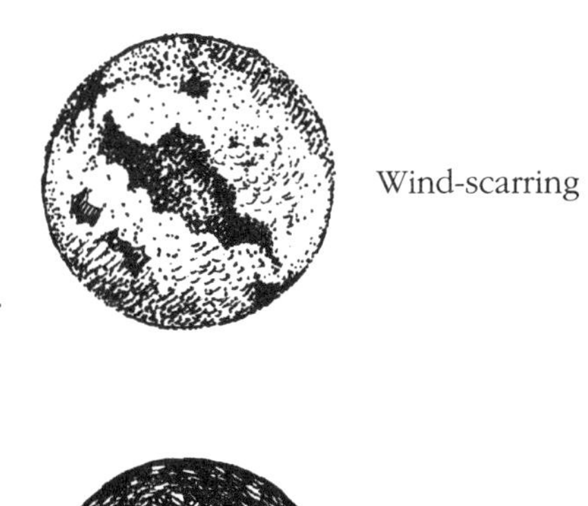


**Fig. 20-2** *Oranges and other citrus fruits also have cosmetic damage.*

First, find a booklet that describes the different types of damage found on apples and citrus fruits. Many booklets contain actual pictures of the damage. Your local library might be able to help, but your best resource is the U.S. Department of Agriculture or the agricultural extension service of your state university. For example, the Cooperative Extension Service at the University of Massachusetts offers a booklet entitled *Integrated Pest Management of Apple Pests in Massachusetts and New England.* The Cooperative Extension Service of Michigan State University offers a booklet entitled *Tree Fruit Insects.*

Study the different types of damage caused by pests. When you have become somewhat familiar with the types of damage found on apples and citrus fruits, go to an organic supermarket and look through the apples for examples of the damage described in the booklet. Find as many different kinds of pest damage as possible. Repeat this procedure with oranges. You might want to buy a few specimens of apples and oranges that have good examples of the different types of damage so that you can draw or take pictures for your report.

When you feel comfortable identifying insect pest damage on both types of fruit, move on to the next part of the project. Find a type of apple (Macintosh, Cortland, Delicious, etc.) that is available at both the organic and regular supermarkets. Do the same for oranges. Go back to the organic supermarket (bring the identification booklet along). Record your observations in a chart similar to TABLE 20-1. Randomly select at least 30 apples from the produce section and check each for the types of insect damage found. Check off the types and amount of damage found on each, if any. When you've finished with the apples, repeat this procedure with oranges. If possible, go to another organic market and repeat this procedure to validate your data. Purchase one or two typical specimens to photograph, and taste later. Note the price per pound.

When you are finished at the organic stores, repeat the entire procedure at a regular supermarket using the same types of produce. Use another data collection table and use at least 30 of each type of fruit. If possible, repeat the process at a second supermarket to validate your results. Once again, purchase one or two typical specimens to photograph, and taste later.

Return to your home or lab to analyze the data. First, weigh each fruit and record the results. Draw or take pictures of the damage for your report. Then, taste each fruit purchased and record any difference or similarity in taste between the two groups. Tally the types and amount of damage found in each group. If you were able to survey more than one organic or regular supermarket, average the results together. Compare the types and amount of damage found on both groups.

**Table 20-1**

| ORGANIC STORE #1 | | | | | | | | | | |
|---|---|---|---|---|---|---|---|---|---|---|
| Apple # | TYPE OF DAMAGE | | | | | | | | | |
| | A | B | C | D | E | F | G | H | I | J |
| 1 | I | II | | | | | | III | | 𝍸 |
| 2 | | | | | | | | | | |
| 3 | | | | | | | | | | |
| 4 | | | | | | | | | | |
| 5 | | | | | | | | | | |
| 6 | | | | | | | | | | |
| 7 | | | | | | | | | | |
| 8 | | | | | | | | | | |
| 9 | | | | | | | | | | |
| 10 | | | | | | | | | | |
| 11 | | | | | | | | | | |
| 12 | | | | | | | | | | |
| 13 | | | | | | | | | | |
| 14 | | | | | | | | | | |
| 15 | | | | | | | | | | |
| 16 | | | | | | | | | | |
| 17 | | | | | | | | | | |
| 18 | | | | | | | | | | |
| 19 | | | | | | | | | | |
| 20 | | | | | | | | | | |
| 21 | | | | | | | | | | |
| 22 | | | | | | | | | | |
| 23 | | | | | | | | | | |
| 24 | | | | | | | | | | |
| 25 | | | | | | | | | | |
| 26 | | | | | | | | | | |
| 27 | | | | | | | | | | |
| 28 | | | | | | | | | | |
| 29 | | | | | | | | | | |
| 30 | | | | | | | | | | |
| Totals | | | | | | | | III | | 𝍸 |

## Analysis

How much damage was found on the organic versus the regular produce? Was the difference substantial? Calculate the percentage of each fruit damaged by each type of pest for both types of supermarkets and record the data in a table similar to TABLE 20-2. Is the difference significant? How does the price differ? Was there any difference in taste between the two groups? Is the difference purely cosmetic?

**Table 20-2**

| | TYPE OF DAMAGE FOUND (% of total) | | | | | | | | | |
|---|---|---|---|---|---|---|---|---|---|---|
| Apples | A | % | B | % | C | % | D | % | E | % |
| Supermarket (30) | 5 | 17% | | | | | | | | |
| Organic foods (30) | 15 | 50% | | | | | | | | |

## Going Further

- Purchase a large sampling of organic and regular supermarket produce. Devise a taste test and a questionnaire. Gather a group of friends and have them sample both types of fruit and then answer your questions. How do most people feel about the differences between the two groups? Do their answers differ if you inform them about the environmental problems caused by purchasing fruit from the regular supermarket? Do they differ when informed about price?
- Much of the damaged organic produce is discarded prior to sale. See if you can run a similar experiment but use produce directly from harvest.

## Suggested Research

- Investigate what the term *organic* really means when it refers to our food. Are there any local, state, or federal laws that regulate the use of the phrase "organically grown"? Research this in your state. Look into California's organic labeling law (even if you don't live in that state).
- Look into what is meant by "organic beef."

# 21

# Insect repellents
## Caveat emptor

*(Do some insect repellents work better than others, and do some health and beauty products impair their effectiveness?)*

Many bloodsucking insects belong to the order *Diptera*, which are the true flies and which possess only one pair of wings instead of the usual two pairs found on other insects. (See FIG. 21-1.) Bloodsucking flies include *mosquitoes*, *black flies*, *greenheads*, *deer flies*, *horse flies* and *no-see-ums*, to name a few. Blood is an excellent source of protein and is often vital to a female insect's ability to produce eggs.

**Fig. 21-1** *Members of the order Diptera such as this mosquito only have one pair of wings instead of the usual two pairs.*

Bloodsucking insects bother more than just humans. They're the pests of other mammals as well as birds, reptiles, and a few amphibians (such as frogs). Many blood-feeding insects find their meal by sensing carbon dioxide that is exhaled after each breath an organism takes.

## Project Overview

On the market are numerous insect sprays that claim to protect us from the onslaught of the bloodsuckers. There are many other products—such as perfumes, shampoos, hair sprays, etc.—that may attract or offend insects.

Do some insect repellents work better than others? Do some health and beauty products, such as perfumes or hair sprays, attract or repel bloodsucking insects? Might some natural products such as garlic or lemon juice attract or repel these insects? Do some of these products counteract one another or is there a synergistic effect when more than one of these products are worn together? To answer these and any other questions your research leads you to begin your literature search and formulate your hypotheses.

## Materials

- Five or more plastic-coated paper plates (about 8 inches in diameter, one for each product to be tested)
- Talcum powder
- TangleFoot or similar insect-trapping substance (available at garden centers)
- Bookends, or anything that will prop up the paper plates on a table (one for each product tested)
- At least 10 large paper clips (1½ inches)
- Outdoor table

- Dry ice, 3 to 5 lbs. (CAUTION: Always use tongs and gloves when handling dry ice.)
- Medium-size plastic bowls to hold the dry ice (one for each product tested)
- A variety of spray-on insect repellents (use at least one repellent that has DEET as the active ingredient.)
- Some miscellaneous products, such as hair spray, perfume, shampoo, garlic, lemon juice, (or anything you think might attract or repel insects).
- Plastic spray bottle (optional)
- Blender (optional)

## Procedures

Late spring until early fall is the best time to perform this project. Label the back of one plate "control." Label the back of each of the other plates with the name of each product to be tested.

Spread TangleFoot over each plate with the brush. (Follow the instructions for application.) Cover the same area on each plate. This substance traps any insects that come in contact with the plates.

When the mix has been applied to the plates, let it sit for about 20 minutes to stabilize. Then, gently spray each plate with the appropriate product—except for the control, which has nothing added. A light spray is all you need. Don't overspray or you will impair the ability of the substance to trap insects. For products that don't normally come in a spray, such as shampoo, create your own mixtures with water and place them into a plastic spray bottle like the kind used to mist plants. For others, such as garlic, use a blender with some water to create a spray.

When the products have been applied to the plates, attach each plate to a bookend with two large paper clips (you can use anything that will hold the plate up in a vertical position). Repeat this procedure for each plate. Place each propped-up plate on a table outside. Separate them equally and place them as far away from each other as possible. For best results, choose an area where you have seen or have been bitten by such insects as mosquitos.

Use tongs to place dry ice into each of the bowls (*Do not use your hands*). Position a bowl in front of each plate. Keep them all evenly spaced and facing the same direction. Position the dry ice so that as it melts, the plates are exposed to equal amounts of gaseous carbon dioxide released by the ice. (The carbon dioxide will attract biting insects.) If there is a slight breeze, be sure that the wind blows from the bowls toward the plates.

Midafternoon to early evening is the best time to run this experiment, and you'll get the best results on a windless day. (A windy day might blow the plates away and will cause an uneven distribution of the carbon-dioxide flow over the plates.)

Check the plates every hour and especially around dusk. Leave the setup in place overnight and check it again the next morning. During each observation, count the number of insects stuck to each plate and record the results. Be sure to also observe the presence of insects flying around each plate even if they haven't landed. Include the time of your observations in your notes. Once you have dismantled the setup, try to identify as many of the trapped insects as possible using your insect guide.

## Analysis

Which plate had the most insects? How did the repellents appear to work individually and in relation to each other? Record your data in TABLE 21-1 or one similar to it. Cross-check the results with the active ingredients of each insect repellent. Were the results different if the active ingredient was different? Did the concentration of the active ingredient make a difference? Can you determine which active ingredient and which concentration of that ingredient worked the best? Did some of the sprays work longer than others? Did the health and beauty products attract or repel insects?

## Going Further

- Try applying one of the products that attracted insects on the same plate with the most effective insect repellent. Create a chart, similar to TABLE 21-2, that depicts the relative success rate of an insect repellent when used with a variety of other products that people often use, such as hair spray. Do some products counteract the effectiveness of an insect repellent when we wear them?
- You can also continue this experiment to see if there is a certain time of day when biting flies are most common. Run this test (without repellent) at different times. Do you get different numbers of insects at different times? Do you get different kinds of insects at different times?

## Suggested Research

- There are many "natural" insect repellents sold that may or may not work as well as those already tested. Some of these repellents are sold through natural food stores or through mail-order companies. Contact these outlets and ask for literature about these products and claims of their effectiveness.
- Read more about how insects find a host to feed on. What senses do they use and what parts of their body are used for this sensory perception?

**Table 21-1**

| | Number of insects caught (minutes) | | | | | | | | | | | | |
|---|---|---|---|---|---|---|---|---|---|---|---|---|---|
| Product | 10 | 20 | 30 | 40 | 50 | 60 | 70 | 80 | 90 | 100 | 110 | 120 | 24 hrs. |
| A | I | III | ~~IIII~~ | | | | | | | | | | |
| B | 0 | 0 | | | | | | | | | | | |
| C | 0 | II | | | | | | | | | | | |
| D | ~~IIII~~ | ~~IIII~~ II | | | | | | | | | | | |
| E | II | III | | | | | | | | | | | |
| F | I | I | | | | | | | | | | | |

**Table 21-2**

| | Plus | Number of Insects trapped | | | | | | | | | | | |
|---|---|---|---|---|---|---|---|---|---|---|---|---|---|
| Product | Product | 10 | 20 | 30 | 40 | 50 | 60 | 70 | 80 | 90 | 110 | 110 | 120 |
| A | B | | | | | | | | | | | | |
| | C | | | | | | | | | | | | |
| | D | | | | | | | | | | | | |
| | E | | | | | | | | | | | | |
| | F | | | | | | | | | | | | |
| B | A | | | | | | | | | | | | |
| | C | | | | | | | | | | | | |
| | D | | | | | | | | | | | | |
| | E | | | | | | | | | | | | |
| | F | | | | | | | | | | | | |

# 22

# Sugar
## Au naturel

*(How much sugar is in honey?)*

When you see a honeybee buzzing about, apparently preoccupied with every nook and cranny of a flower, it is probably searching for nectar. When collected, bees store this nectar in a part of their digestive system called the crop. When their crop is full, bees return to their hive and regurgitate the substance into a cell in the honeycomb. (See FIG. 22-1.) Other bees do the same until the cell is full. Over time, the water in this solution evaporates, and what remains is honey.

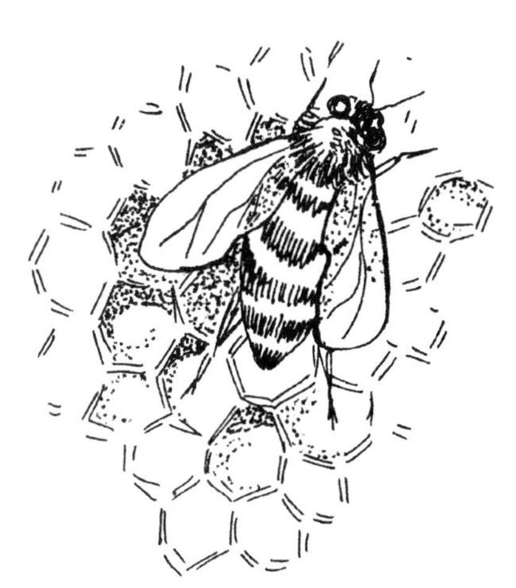

**Fig. 22-1** *Honeybees deposit nectar into the honeycomb.*

## Project Overview

Insects are vital parts of many food webs and are an important food source for many other organisms. In some parts of the world, even humans eat insects, such as the hordes of locusts that are baked and eaten in some portions of northern Africa. Insects are rich in protein and contain many other nutrients.

Although most people would not eat insects per se, many do consume a product made by insects—honey, which is often used as a natural sweetener to replace processed sugars. People use honey in place of sugar while baking or cooking; and they use it with cereal, tea, or coffee. (See FIG. 22-2.)

**Fig. 22-2** *People often use honey as a natural replacement for refined sugars.*

Honey contains sugar. If we use honey as a natural replacement for processed sugar, it would be beneficial to know how much sugar is actually in honey. How does the sugar content of honey compare to that of an equal volume of sugar? (This project involves no live insects.) To answer these and any other questions that your research leads you to, begin your literature search and formulate your hypotheses.

## Materials

- Honey
- Fructose or glucose sugar (You can purchase these at a grocery store. Do not use regular cane sugar!)
- ⅛-, ¼-, ½-, and 1-teaspoon measurers
- Marker that writes on glass
- Eyedropper
- Quantitative Benedict solution (available at a supply house or possibly your school lab)
- Small pot (approx. 1 pint)
- Test-tube holder or an oven mitt
- Five similar test tubes (About 6" long & ⅝" in diameter)
- Test-tube stirrer
- Test-tube rack
- Ruler

## Procedures

This experiment can be done at any time of the year. You will create a series of sugar solutions and will test the sugar content of each with Benedict solution. You'll also test a solution containing honey. Use the ruler to measure ½ inch up from the bottom of each test tube and draw a line on the tube at this level. Measure ½ teaspoon of sugar (fructose or glucose, not sucrose) and pour it into one of the test tubes. Label this tube "½ teaspoon." Using the eyedropper, add enough water to bring the level to the ½-inch line on the tube. Swirl the tube until the sugar is dissolved. Place it in the test-tube rack. (See FIG. 22-3.)

In another tube, measure in ¼ teaspoon of sugar, and label this tube "¼ teaspoon." Again, add enough water to bring the level to the ½-inch

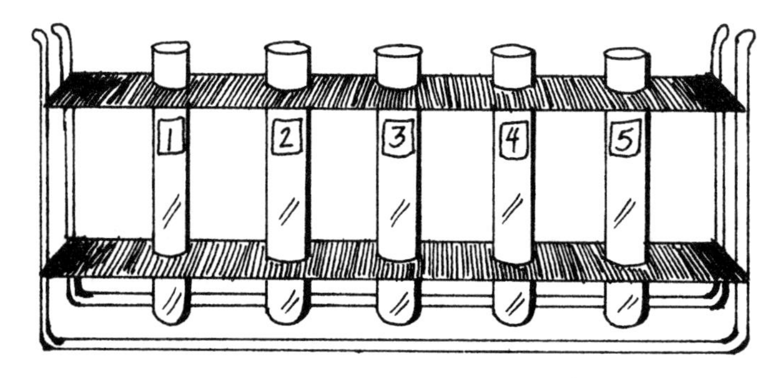


**Fig. 22-3** *Place the five test tubes containing the five solutions into a test tube rack.*

line on the tube; swirl until dissolved and place in the rack. In a third tube, add ⅛ teaspoon of sugar, label accordingly, add water up to the line, swirl until dissolved, and place in the rack. In a fourth tube, add water up to the line and label this tube "blank." Finally, in a fifth tube, add ½ teaspoon of honey and label the tube "honey." Add water to bring the fluid level up to the ½-inch line and swirl to dissolve (you might have to stir the contents to get it to dissolve).

When all the tubes are mixed and labeled, add 2½ teaspoons of Benedict solution to each tube. Now, boil water in a small pot. The water should only be about 2 inches high in the pan. Place the test tubes in the boiling water for 2 to 3 minutes as shown in FIG. 22-4. The tops of the test tubes must be well above the water's surface. When finished heating, *use an oven mitt or a test-tube holder* to remove the tubes from the pot and place them back in the rack.

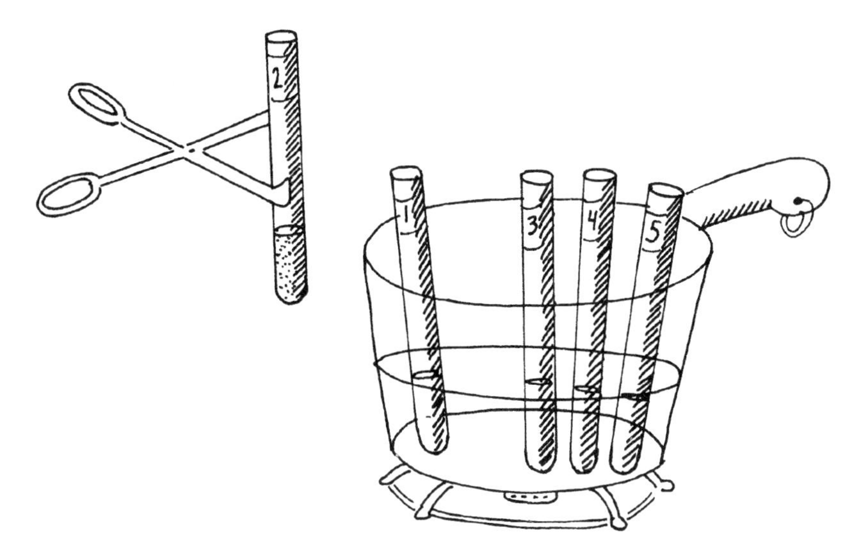


**Fig. 22-4** *Place all the test tubes into a small pan containing boiling water.*

What are the colors of the solutions in each test tube? (Benedict solution turns from yellow to a brownish yellow depending on how much sugar is present. The more sugar the darker the color.) After the tubes cool, look for precipitate (solid material that sinks to the bottom of the tube). Do all the tubes have precipitate? (More precipitate means more sugar was present.) How does the reaction in the tube containing honey compare with the sugar solutions? Do the results of the test tube containing honey fall somewhere between two of the sugar solutions?

Now try to fine-tune this experiment to get the same reaction between a specific amount of sugar and the same volume of honey. To do this, select the sugar concentration that had results closest to the honey results. Repeat the experiment using various concentrations of sugar each time until the color and precipitate are the same as in the test tube with honey.

## Analysis

How much sugar is in the test tube containing honey? Now that you know the concentration of sugar (in water) that matches an equal volume of honey (in water), can you determine how much sugar is in one teaspoon of honey? During your literature search be sure to read about the different kinds of sugars that exist—for example, which ones are present in honey and which are found in common sugars purchased in a supermarket.

## Going Further

Find out what other methods, besides the one with Benedict solution, can be used to study the chemical makeup of honey. Can you devise experiments that determine both the amount and the type of sugar present?

## Suggested Research

- Most people think that honey is healthier than processed sugar. Can you find proof of this or are these claims speculative?
- Read more about how honey is mass-produced by companies that bottle and sell it. Are any preservatives or other additives used in honey? If so, why? And if not, why not?

# 23

# Insects
## Living with them

*(What insects exist in your home,*
*and what are their living preferences?)*

Insects have shared our homes ever since we started building them. Insects that move in with us are considered at the very least a nuisance and at worst a threat. Many species of insects have actually evolved to adapt to our dwellings, making our homes their homes as well.

Many insects live in close association with humans. Some insects, such as lice and fleas, have established parasitic relationships with us. Many simply find our homes perfect places to live and eat—including silverfish, booklice, termites, and flour beetles. The most well-known resident, however, is the cockroach. Cockroaches are found in almost every place where people are found. They can be difficult to get rid of since they can eat any number of things and make their home out of almost anything.

People spend a great deal of time trying to rid their dwellings of these insect invaders. Indoor insect traps and sprays are an annual multimillion dollar industry. (See FIG. 23-1.) Before any form of insect control can be found, whether it be traditional or some alternative method, scientific research must be performed. This includes first surveying what insects exist in a structure and identifying how they survive.

## Project Overview

Before an indoor insect can be controlled, it must be identified and its natural history studied in order to determine such things as what it feeds

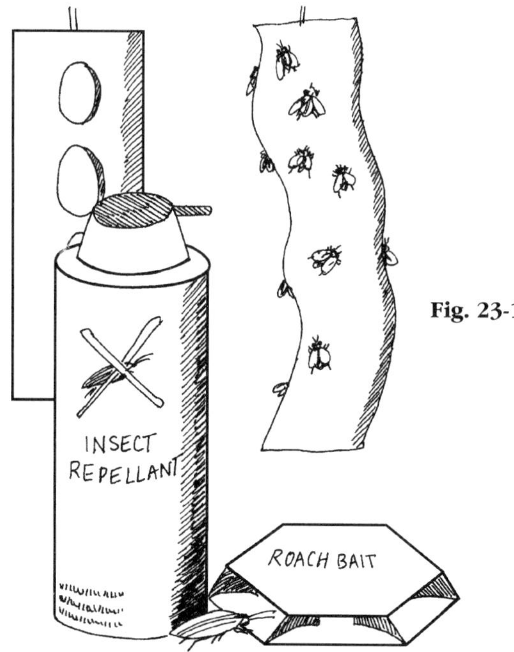


**Fig. 23-1** *Controlling insect pests in the home is a multimillion-dollar business.*

on and what kind of material it lives on or in. In this project you will first collect insects from various buildings (or parts of the same building) in order to identify what type of insect is the most common. You'll then create an insect survey collection that demonstrates the types and relative quantities of each insect. A survey collection contains all the insects found in a certain area or habitat. In this case, the habitat is a building. When such a survey is made, the collection can be studied for the purpose of drawing conclusions about the insect populations in that habitat.

After the most common insect has been identified, you will determine the insect's preferred bedding material. This information could enable you to determine how to control the insect—for example, by spraying chemicals on material that will likely be used by the insect for bedding, or by simply removing the bedding material. Additionally, a test will be performed to see what happens if an insect's favorite bedding is not available.

Although you can perform the second part of this project with any insect that you identify in the first part, the "Procedure" section that follows gives instructions specifically for the cockroach. You can adapt this experiment, however, to any insect by adjusting the bedding materials used.

What kinds of insects do you think are most common in the building that you plan to study? Do different kinds of insects live in different parts of the building? What is their favorite bedding material? To answer these and any other questions that your research leads you to, begin your literature search and formulate your hypotheses.

## Materials

The first part of the experiment (collecting insects) requires the following:

- 10 wide-mouthed jars (32-ounce mayonnaise jars, or quart mason jars)
- Vaseline
- Coffee filters (no. 4 size, cone-shaped)
- Newspaper
- Plastic masking tape
- Oats
- Banana slices
- Honey
- Yeast cakes

- Activating fluid, such as ethyl acetate (You can use nail-polish remover)
- Cotton balls
- Small plastic bags (to hold the insects that were in the jars)
- Plastic bags (to fit over mouth of jar)
- Rubber bands

The second part of the experiment, in which you test to find an insect's favorite bedding material, requires the following (this section assumes that you are testing for cockroaches, but you should also research to see what bedding materials can be used for other types of insects).

- Five cockroaches (or other insect identified as the most common in the first part of this experiment). (These can be specimens collected from the first part of the experiment or they can be ordered from a supply house.)
- Two shoe boxes
- Nylon netting material
- Plastic packing tape
- Cardboard (about 2 feet square)
- Newspapers
- Scissors
- Plastic sheet (such as a plastic grocery or garbage bag)
- Sand
- Four pieces of dry dog food

## Procedures

Since both parts of this experiment are done inside, they can be performed at any time of year. In the first part, you will make 10 insect traps that will be placed in various locations within a building—for example, your home, your school, an office building, or just about any other indoor facility. Be sure to get permission from the appropriate people before setting any traps in the building (speak with your sponsor about who to get this permission from). If you prefer, you can collect specimens in a variety of buildings.

The traps that you build will allow insects to crawl into the jars, but will make it difficult for them to escape. (See FIG. 23-2.) To make a trap, put a layer of Vaseline on the upper inside edge of each jar, about ½ inch from the top. The Vaseline should be approximately ⅛ inch thick and ¼ inch wide. This keeps insects from crawling back out of the jar. Cut a 1-inch hole in the tip of each coffee filter. Put each filter two-third of the way into the mouth of each jar and fold the upper edge of each filter over the outside edge of each jar, as shown in FIG. 23-2. This, too, helps the insects get in but makes it difficult for them to get out.

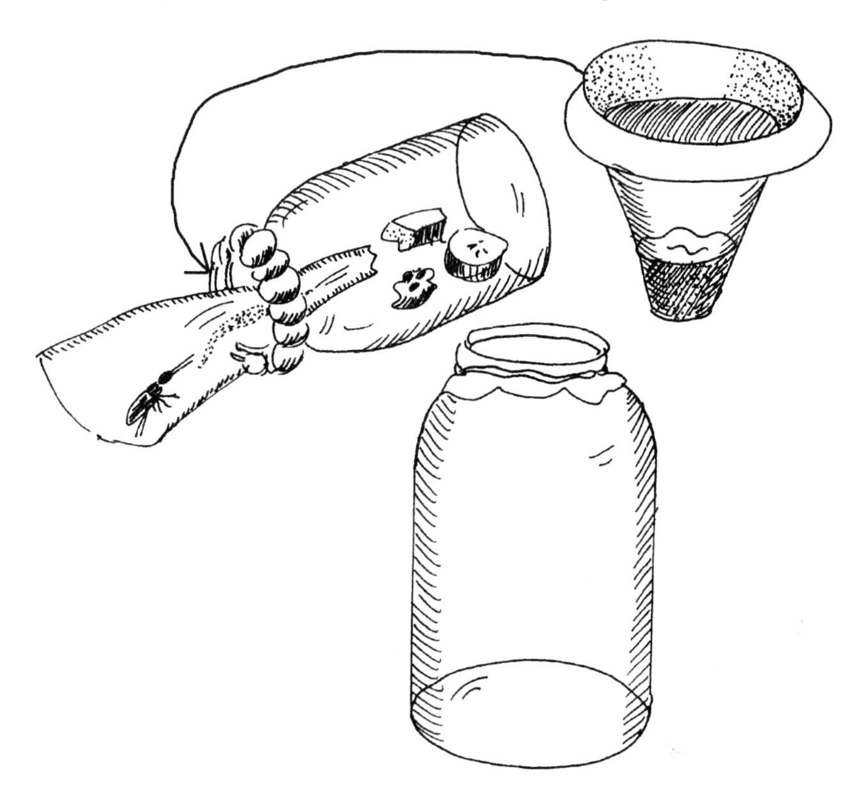

**Fig. 23-2** *You will build a series of traps that will attract a wide range of indoor insect pests.*

Now, make some ramps so that it's easy for crawling insects to enter each jar. For each ramp, cut a 2-inch by 4-inch piece of newspaper. Tape the newspaper to the edge of the cone (which is inserted into the jar) to make a ramp as shown in FIG. 23-2. Now each trap is complete and only needs to be baited. Just before using a trap, remove the paper cone and attached ramp so that you can bait the trap. Place a slice of banana, a spoonful of oats, a drop of honey, and half of a moistened yeast cake in the jar. Put back the cone and the ramp—and now you have a baited trap. This wide range of foods will attract just about any kind of insect living in the building.

Place the traps in a variety of places within the building surveyed. Basements, kitchens, bathrooms, storage areas, and garages are all good trap sites. The traps should be placed against a wall, in an out-of-the-way spot. Dark, damp places are usually good for catching insects. Record the date and time that you set the traps and the location of each so that you can find them in a few days. The traps should be left in place for 3 to 5 days.

When enough time has passed, return to each location. If a trap has no insects, you can use the trap again by refreshing the bait and by making sure that the Vaseline layer is still sufficient; but find a new location for that particular trap (and document the fact that no insects were found at that location). For the traps that contain insects, do the following: (1) take a cotton ball and dampen it preferably with ethyl acetate or else nail-polish remover; (2) take the cone off the trap and drop in the cotton; and (3) cover the trap with a plastic bag and use a rubber band to hold it in place. (See FIG. 23-3.) Continue collecting all the jars in this manner. (If a trap is found in disarray and the bait is gone, mice are probably the

**Fig. 23-3** *Insert the cotton ball containing ethyl acetate into the trap and cover with a plastic bag.*

culprits. Reset the trap at another location since the mice will most likely return.)

After the trapped insects have been in the jars with the cotton for about 2 hours, dump them out into separate plastic bags. Mark each bag with the building and trap location as well as the date and time when the trap was emptied. After you have collected all the traps and emptied the contents of each into separate bags, count the insects collected at each trap. Use your insect field guide to identify as many as possible. Be sure to keep track of what insects were found and where they were found.

After completing the first part of the experiment, identify which insect was found most often in the building surveyed. In the second part of the experiment, you will determine what this insect prefers for bedding material. This will help you understand where these insects are living in the building and what control measures might be taken.

You will make a four-section cage with shoe boxes as shown in FIG. 23-4. Cut each box in half (width-wise). Tape all four halves together in a cross shape as shown in FIG. 23-4. On the inside of the box, tape any crevices and holes shut. Tape a piece of cardboard to act as a floor for the middle section of the cross, since the cage will be missing a floor. Tape up all the edges so that insects cannot escape.

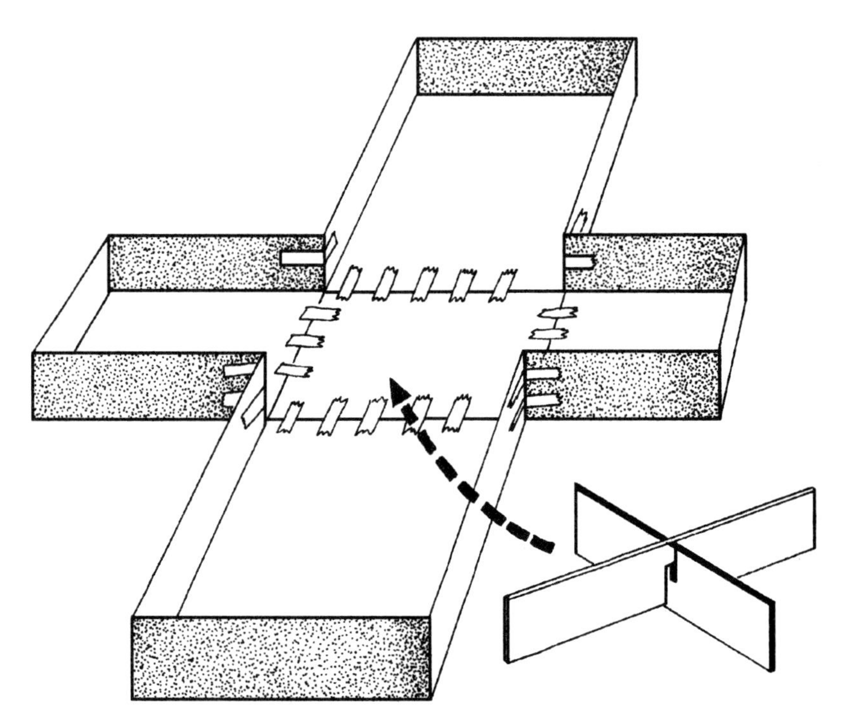

**Fig. 23-4** *This apparatus, made of two shoe boxes, simulates four separate environments for an insect pest.*

When the box is completed, fill each section with a different bedding material. If you are using cockroaches, cut some newspaper strips and place them in one of the four portions of the box (the strips should be loosely packed). Fill another portion with an inch of dry sand. Fill another portion with damp newspapers (not dripping wet), placed on top of one of the plastic bags to keep the box from getting wet. Leave the fourth and last portion of the box empty, so that it just has the cardboard floor of the shoe box.

Now you'll supply food for the insects to eat throughout the duration of the experiment. Slightly moisten the dog food by putting five drops of water on each of four pieces. Place one piece in each of the four portions of the box. Cover the top of each of the four portions of the box with nylon material (tape the material to the boxes to hold it on). Do not cover the center portion of the box yet, but have a piece of nylon cut and ready. The center portion must be left open until the insects have been placed inside.

Drop five cockroaches (or other type of insect) into the center of the box. Quickly cover the center of the box with nylon netting and tape it shut (be sure you have taped all the holes). Leave the cage undisturbed for 1 week, in a darkened room (or, cover with newspapers).

Before the week has passed, for the final stage of the experiment, you must make a barrier that keeps the insects from moving from one portion of the box to another. This barrier will let you see where the insects have been living for the past week. To create this barrier, cut two pieces of cardboard to fit kitty-cornered into the center of the box. Make a center cut halfway up the middle of each piece so they can be fitted together to form an X.

At the end of the week, carefully and quickly remove the center nylon piece and place the cardboard X in place, so that the insects are trapped in the portions of the box where they have been residing. (See FIG. 23-5.) Tape the cross in place and tape the nylon back on the top of the cross so that the insects cannot crawl out.

Now count all the insects in each section of the box (you might need to remove the nylon cover for each section as you examine it). Note how many insects are found in each section.

## Analysis

What kinds of insects did you collect in the first part of the experiment? Where did you collect the most insects? Did you get different kinds of insects at different sites? How many different insect orders were found? Were many different species of insects found to inhabit the building, or were there large numbers of only a few species? Compile a survey collection of the insects captured that not only shows the types of insects caught but also the relative numbers of each. (See FIG. 23-6.)

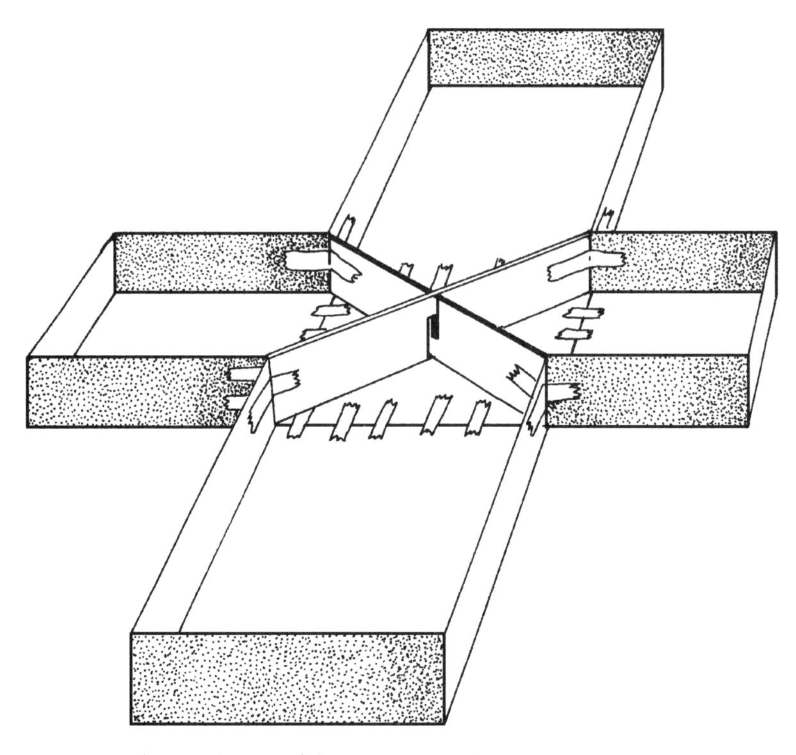

**Fig. 23-5** *At the conclusion of the experiment this divider is placed in the middle of the apparatus to keep the insects in their respective corners.*

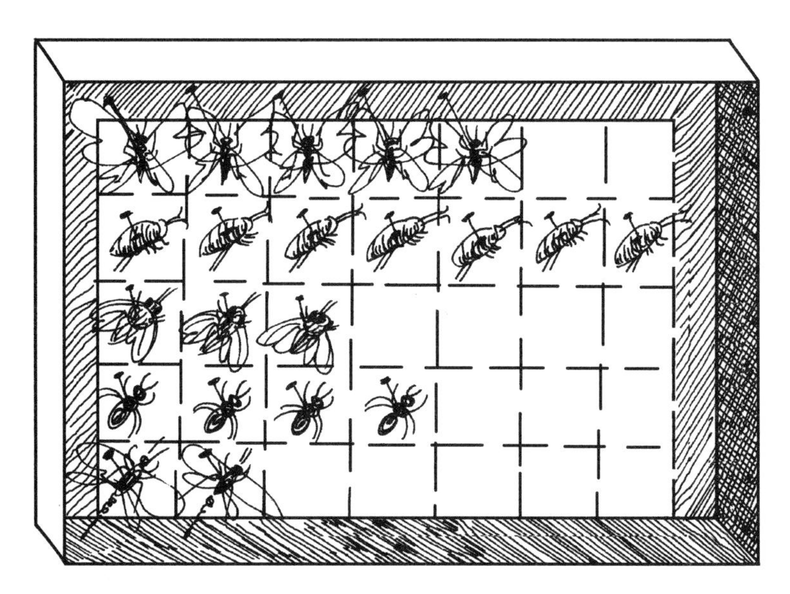

**Fig. 23-6** *An insect collection can show not only the types of insects collected, but also the relative numbers of each.*

In the second part of the project, did one portion of the box have more insects than any other? Which bedding materials did they like the most, and which the least? Can you conclude that cockroaches (or other domestic insects) prefer a particular bedding material more than any other for habitation? If so, why do you think one is preferable to another? How might this information be used by persons who are trying to control these insects?

## Going Further

- Repeat this experiment, but eliminate the material that the insects preferred in the first part of this project. Do the insects adapt to another bedding material? What happens if they cannot find a suitable material?
- Continue this experiment to determine an insect's preferred food. Place different foods (dog food, fruit, sugar, or cereal) in each portion of the cross. Collect the roaches (or other type of insect) one week later using the same techniques described in the original experiment. Which food do the insects prefer?

## Suggested Research

- How might what you have learned about insects' favorite bedding and food be used to control pests?
- Read more about domestic insects (insects that live with humans). How have insects affected human history? What have we done over the centuries that encourage insects to live with us?

# Appendix A

# Using metrics

Most science fairs require that the metric system be used for all measurements. Meters and grams, which are based on powers of 10, are actually far easier to use during your experimentation than feet and pounds. You can convert any English units into metric units if need be, but it is easier to simply begin with metric units. If you are using school equipment, make sure, for example, that flasks or cylinders are marked with metric units. If you are purchasing your glassware (or plastic ware) be sure to order metric markings.

Conversions from English units to metric units (along with their abbreviations) are given below (all conversions are approximations).

*Length*:
one inch (in.) = 2.54 centimeters (cm)
one foot (ft.) = 30 cm.
one yard (yd.) = .90 meters (m)
one mile (mi.) = 1.6 kilometers (km)

*Volume*:
one teaspoon (tsp.) = 5 milliliters (ml)
one tablespoon (tbsp.) = 15 ml
one fluid ounce (fl. oz.) = 30 ml
one cup (c.) = .24 liters (l)
one pint (pt.) = .47 l
one quart (qt.) = .95 l
one gallon (gal.) = 3.80 l

*Mass*:
one ounce (oz.) = 28.00 grams (g)
one pound (lb.) = .45 kilograms (kg)

*Temperature*:
32 degrees Fahrenheit (F) = 0 degrees Celsius (C)
212 degrees F = 100 degrees C

To convert F to C use:

$$\frac{(F - 32) \times 5}{9}$$

To convert C to F use:

$$\frac{(C + 32)}{5/9}$$

# Appendix B
# Sources

## Suggested Reading

If you are new to science fairs, here are a few good books to read. They cover all aspects of entering a science fair.

Tocci, Salvatore. 1986. *How to Do a Science Fair Project.* New York: Franklin Watts.

Iritz, Maxine. 1987. *Science Fair: Developing a Successful and Fun Project.* Blue Ridge Summit, PA: TAB Books.

Bombaugh, Ruth. 1990. *Science Fair Success.* Hillside, NJ: Enslow Publishers.

The following books can all be used for additional ideas regarding science fair projects. Although not specifically about insects, many of the books involve insects in some way or another.

Irtiz, Maxine. 1991. *Blue-Ribbon Science Fair Projects.* Blue Ridge Summit, PA: TAB Books.

Bochinski, Julianne. 1991. *The Complete Handbook of Science Fair Projects.* New York: Wiley and Sons.

Berman, William. 1986. *Exploring with Probe and Scalpel. How to Dissect: Special Projects for Advanced Studies.* New York: Prentice-Hall.

Witherspoon, James D. 1993. *From Field to Lab: 200 Life Science Experiments for the Amateur Biologist.* Blue Ridge Summit, PA: TAB Books.

Gutnik, Martin J. 1991. *Experiments That Explore Oil Spills.* Brookfield, CT: Millbrook Press.

Gutnik, Martin J. 1992. *Experiments That Explore Recycling.* Brookfield, CT: Millbrook Press.

The following are insect field guides to help you identify insects collected for the projects in this book.

Borror, D. J., and and R. E. White. 1970. *A Field Guide to the Insects of America North of Mexico.* Boston: Houghton Mifflin.

Bland, R. G., and H. E. Jaques. 1978. *How to Know the Insects.* Dubuque, IA: W. C. Brown.

Zim, H. S., and C. Cottam. 1951. *Insects: A Guide to Familiar American Insects.* New York: Simon and Schuster.

Arnett, R., and R. Jacques. 1981. *Simon and Schuster's Guide to Insects.* New York: Simon and Schuster.

Audubon Society and L. Milne. 1980. *The Audubon Society Field Guide to North American Insects and Spiders.* New York: Knopf.

Borror, D. J., and D. M. DeLong. 1970. *A Field Guide to the Insects.* Peterson Field Guide. Boston: Houghton Mifflin.

Dashefsky, H. S., and J. G. Stoffolano. 1977. *A Tutorial Guide to the Insect Orders.* Minneapolis: Burgess.

For field guides that specialize in aquatic insects try one of these.

Lehmkuhl, D. M. 1979. *How to Know the Aquatic Insects.* Dubuque, IA: W.C. Brown.

McCafferty, W. P. 1982. *Aquatic Entomology.* New York: Jones and Bartlett.

For a colorful overview of all aspects of entomology, try this excellent book.

Imes, Rick. 1992. *The Practical Entomologist.* New York: Simon and Schuster.

For an in-depth, comprehensive reference book on all aspects of entomology, try the following excellent college textbook.

Romoser, William S., and John G. Stoffolano. 1993. *The Science of Entomology.* Dubuque, IA: W. C. Brown.

For information about the International Science and Engineering Fairs and valuable information about adult sponsorship, write to the Science Service at 1719 N Street, N.W., Washington, DC 20036, or call (202) 785-2255.

## Scientific Supply Houses

You can order equipment, supplies, and live specimens for projects in this book from these companies. For your convenience a list of catalog order numbers from Ward's Scientific Supply is listed on the following pages, categorized by project number.

Blue Spruce Biological Supply Company
221 South Street
Castle Rock, CO 80104
(800) 621-8385

The Carolina Biological Supply Company
2700 York Road
Burlington, NC 27215
Eastern US: 800-334-5551
Western US: 800-547-1733

Connecticut Valley Biological
82 Valley Road
P.O. Box 326
Southampton, MA 01073

Fisher Scientific
4901 W. LeMoyne Street
Chicago, IL 60651
800-955-1177

Frey Scientific Company
905 Hickory Lane
P.O. Box 8101
Mansfield, OH 44901
(800) 225-FREY

Nasco
901 Janesville Avenue
P.O. Box 901
Fort Atkinson, WI 53538
(800) 558-9595

Nebraska Scientific
3823 Leavenworth Street
Omaha, NE 68105
(800) 228-7117

Powell Laboratories Division
19355 McLoughlin Boulevard
Gladstone, OR 97027
(800) 547-1733

Sargent-Welch Scientific Company
P.O. Box 1026
Skokie, IL 60076

Southern Biological Supply Company
P.O. Box 368
McKenzie, TN 38201
(800) 748-8735

Ward's Natural Science Establishment, Inc.
5100 West Henrietta Road
Rochester, NY 14692
(800) 962-2660
*or*
815 Fiero Lane
P.O. Box 5010
San Luis Obispo, CA 93403
(800) 872-7289

## Ward's Natural Science Catalog Numbers

Many of the items shown in the "Materials" section of each project can be purchased from Ward's Natural Science, Inc. Below are some of these items (by chapter) and the catalog number to use for ordering.

Chapter 4
Blow fly larvae (87 W 6356)

Chapter 9
Drosophila culturing kits (18 W 1369)
(Numerous strains of Drosophila are available from Ward's.)

Chapter 11
Insect Sweeping Net (10 W 0550)
Insect Dual Purpose Net (10 W 0510)

Chapter 12
Killing jars (10 W 0250)
Activating Fluid (10 W 0187)
Collecting Aspirator (10 W 0175)

Chapter 13
Humus Soil Testing Kit (36 W 5521)
Berlese Funnel (20 W 8003)

Chapter 14
Aquatic insect collecting net (10 W 0600)

Chapter 16
Blow fly pupae (87 W 6355)

Chapter 17
Harvester ants (87 W 6950)
Ant-colony chamber (14 W 7520)

Chapter 18
Same as Chapter 12

Chapter 22
Quantitative Benedict solution (37 W 0704)

Chapter 23
Cockroaches (87 W 6150)

## Entomological Organizations

If you are interested in insects, consider joining or receiving literature from one of these organizations.

Xerces Society
Department of Zoology
University of Wyoming
Laramie, WY 82071

Young Entomologists' Society Inc.
1915 Peggy Place
Lansing, MI 48910-2553
(517) 887-0499

The Amateur Entomological Society
137 Gleneldon Road
Streatham, London SW 16
UK

If your interests are specifically about butterflies and moths, contact one of these groups.

Lepidoptera Research Foundation
Santa Barbara Museum of Natural History
2559 Puesta del Sol Road
Santa Barbara, CA 93105

Lepidopterists's Society
Department of Biology
University of Louisville
Louisville, KY 40208

# Glossary

**abstract** A project summary (usually less than 250 words) that is often required at many fairs.

**backboard** The vertical, self-supporting panel used in your science fair display. The board displays explanations or descriptions of the project, such as a statement of the problem and hypothesis; photos of the experimental setup, organisms, and other important aspects of the project; and analyzed data in the form of charts and tables. Most fairs limit the size of displays.

**biocontrol** The use of organisms to control insect pest populations (also called *biological control*).

**biodiversity** The diversity and implied importance of all organisms on our planet.

**biological categories** Most science fairs categorize projects according to subject area. Awards are usually given in each subject area.

**bioremediation** The use of organisms to clean up waste products, such as oil spills or radioactive material.

**cephalothorax** Two of the three major body regions (head and thorax) combined into one, as in *arachnids* (or spiders).

**collecting aspirator** A suction device for collecting insects (see FIG 1-8.)

**control group** A test group—offering a baseline for comparison—in which no experimental factors or stimuli are introduced.

**coxa** The first segment of an insect leg.

**cuticle** The outermost layer of an insect's exoskeleton, which is produced as a secretion.

**dependent variable** A variable that changes when the experimental (independent) variable changes. For example, in a study of the mortality (death) rate of organisms living in soil exposed to pesticides, the mortality rate would be the dependent variable and the pesticides would be the experimental variable.

**desiccation** The loss of all water.

**detritus** Decomposing organic matter.

**diapause** A period of little or no activity during unfavorable environmental conditions.

**display** The entire science fair exhibit, of which the backboard is a part.

**duff** Decomposed leaf litter.

**exoskeleton** The external supporting and protective structure of an arthropod, such as an insect.

**experimental group** A test group that is subjected to experimental factors or stimuli for the sake of comparison with the control group. The experimental group is the one exposed to the factor being tested—for example, a plot of soil containing organisms exposed to varying amounts of pesticides. (See *experimental variable.*)

**experimental variable** The aspect or factor to be changed for comparison—for example, the amount of pesticide that soaks into the soil. (This factor is also called the independent variable.)

**femur** The third segment of an insect leg.

**gall, insect** An insect-induced growth on a plant that is used for protection and often food by the insect.

**habituation** The gradual reduction of a response to an event (such as a stimulus).

**hypothesis** An educated guess, formulated after thorough research, to be shown true or false through experimentation.

**invertebrates** Organisms with no backbones, such as insects, crustaceans, and mollusks.

**journal** A project notebook, containing all notes on all aspects of a science fair project, from start to finish.

**leaf litter** Partially decomposed leaves, twigs and other plant matter that have recently fallen to the ground forming a ground cover.

**leaf miner** An insect that spends part of its life living in the layers of a leaf. As it feeds it bores or tunnels, mines which also provide protection.

**leaf roller** An insect that curls part of a leaf around its body for protection during an immature stage.

**mechanical control** A method of controlling insects by a mechanical means, such as using oil to block the insects' spiracles.

**metamorphosis** The change in body form during an insect's development.

**observations** A form of qualitative data collection.

**ovipositor** The external female reproductive organ used to lay eggs.

**paedeogenesis** The ability of an immature organism to produce young.

**parasite** An organism that lives in or on another organism (or host) for a portion of its life. The host is not killed in the process.

**parasitoid** An insect that lives in another organism (host) and kills its host during its development.

**parthenogenetic reproduction** The ability to reproduce without a mate (or reproduction without the fertilization of the egg.)

**pathogens** Organisms that cause disease in other organisms.

**population dynamics** The study of populations and the factors that affect them.

**pheromone** A chemical that communicates information to members of the same species.

**predator** An animal that eats other live animals for its nourishment.

**qualitative studies** Experimentation whose data collection involves observations but no numerical results.

**quantitative studies** Experimentation whose data collection involves measurements and numerical results.

**raw data** Any data collected during the course of an experiment that has not been manipulated in any way.

**research** Locating and studying all existing information about a subject (also called a literature search).

**Riker mounts** A flat box used to store and display insects. It has a glass top for viewing and a cotton mat to cushion the specimens.

**scavenger** An organism that consumes dead organic matter.

**scientific method** The basic methodology of all scientific research and experimentation, including the statement of a problem to be solved or the question to be answered to further science, the formulation of a hypothesis and the experimentation to determine if the hypothesis is true or false (experimentation includes collecting and analyzing data and reaching a conclusion).

**smooth data** Raw data that has been studied and then arranged in a way that provides understandable information in the form of graphs and charts that represent totals, averages, and other numerical analysis.

**spiracles** Openings through an insect's exoskeleton that lead to the tracheal system (for carrying oxygen to the insect's cells).

**stimulus** Something that prompts a reaction or a response.

**survey collection** A collection of organisms from a certain habitat or area.

**statistics** The analysis of numerical data.

**sweep net** An insect-collecting net designed to be swept through vegetation for the purpose of collecting large numbers of insects quickly.

**tarsus** The last (fifth) segment of an insect's leg.

**tibia** The fourth segment of an insect's leg.

**tracheal system** A series of tubes throughout an insect's body that carry oxygen to the cells.

**transfer aspirator** A device that allows for the easy transfer of small insects from one area or container to another by the use of suction. (See Fig. 1-9.)

**trochanter** The second segment of an insect's leg.
**variables** A factor that is changed to test a hypothesis.
**vertebrates** Animals with backbones, such as reptiles, amphibians, birds, and mammals.
**wingpads** The immature form of an insect's wing, unable to produce flight.

# Index

## C



## D



## E

## About the Author

The author is an adjunct professor of environmental science at Marymount College in Tarrytown, New York. He is the founder of the Center for Environmental Literacy, which was created to educate the public and business community about environmental topics. He holds a B.S. in biology and an M.S. in entomology and is the author of many books that simplify science and technology.

## Workshops at your school

The author offers hands-on workshops and seminars for teachers on the projects in this book and many other topics. He can be reached at The Center for Environmental Literacy, 383 Main Street, Ridgefield, Connecticut 06877, or call (203) 438-8080.